The Therapeutic Relationship in Analytical Psychology

In *The Therapeutic Relationship in Analytical Psychology: Theory and Practice* Claus Braun presents a thorough exploration of the importance of the therapeutic relationship and explains how to encourage and develop it. Drawing on Braun's decades of clinical experience, the book clearly demonstrates the significance of establishing an intensive and living connection between client and analyst.

The book examines the crucial steps of the psychotherapeutic process, illustrated with a detailed case study that presents the personal development of an analysand through a series of dreams and drawings. Braun connects key concepts in analytical psychology, such as complexes, symbols, archetypes and amplification, with conscious and unconscious processes and the development of the therapeutic relationship during the analytic process. The book also examines why C. G. Jung put such a special emphasis on the therapeutic relationship and explores the ethical demands and social responsibilities of the analyst. Comprehensive and insightful, it skillfully makes the connection between Jung's analytical psychology and practical psychotherapeutic work.

The Therapeutic Relationship in Analytical Psychology will be an essential text for Jungian analysts and psychotherapists in practice and in training and a key reference for academics and students of analytical psychology, psychotherapy and Jungian studies.

Dr. Claus Braun is a clinical psychologist, neurologist, psychiatrist, doctor of psychosomatic medicine and psychotherapy, psychoanalyst and group analyst based in Berlin, Germany. He is a lecturer at the C. G. Jung Institute and at the Institute for Psychotherapy in Berlin, a teaching and control analyst for DGAP and an editorial member of the journal *Analytische Psychologie*.

The Therapeutic Relationship in Analytical Psychology

Theory and Practice

Claus Braun

Routledge
Taylor & Francis Group

LONDON AND NEW YORK

First published 2020
by Routledge
2 Park Square, Milton Park, Abingdon, Oxon OX14 4RN

and by Routledge
52 Vanderbilt Avenue, New York, NY 10017

Routledge is an imprint of the Taylor & Francis Group, an informa business

© 2020 Claus Braun

Originally published in German as *Die therapeutische
Beziehung: Konzept und Praxis in der analytischen Psychologie
C. G. Jungs* W. Kohlhammer

© 2016 W. Kohlhammer GmbH, Stuttgart

British Library Cataloguing-in-Publication Data
A catalogue record for this book is available from the British Library

Library of Congress Cataloging-in-Publication Data
A catalog record has been requested for this book

ISBN: 978-0-367-34710-9 (hbk)
ISBN: 978-0-367-34711-6 (pbk)
ISBN: 978-0-429-32731-5 (ebk)

Typeset in Times New Roman
by codeMantra

'Claus Braun has written a comprehensive, careful and very important book on the therapeutic relationship as understood within analytical psychology. It is an excellent introduction to this topic. I highly recommend it for training in analytical psychology and psychotherapy.'

Marianne Müller, past IAAP President,
Zurich, Switzerland

'Beautifully organised, clear, and yet deeply reflective, Claus Braun's book will delight Jungian clinicians and candidates. In addition, it will also join the growing body of "crossover" works that take the post-Jungian clinical message to non-Jungian readers who may not have kept up with developments in analytical psychology. I wholeheartedly welcome the fact that a volume already acclaimed in the German-speaking world is now available to an Anglophone audience.'

Professor Andrew Samuels, author of
Jung and the Post-Jungians

'Years ago, when I was still a training candidate, I always wished for such a book, one in which form, theory and practice were vividly woven into one another, from where a deeper understanding of what was special, specifically Jungian, in our work could be gained.

By choosing this title, Braun signals his position and his intentions: Braun thinks from the point of view of relationships, and thus the reader feels equally invited to participate in dialogue in current analytic discourses both from the German and Anglo-American linguistic areas and in the extensive history of treatment presented by Braun with cautious systematics.

Braun's intention is to make visible the viability and topicality of Jungian basic concepts in a theoretical discourse. In the prism of the central concept of relationship broken, Jung's essential concepts such as complex, shadow, dream, archetype, symbol, alchemy, amplification, ego and self, transformation, individuation, transmission and countertransference gain a depth of focus that make Braun's approach to representation convincing: in the light of relationship, the terms gain dynamics, they become readable as process figures linked or connectable with each other.

The last third of the book is a case history presented in detail – from the first encounter to farewell. His closing words are a gently but clearly formulated ethical appeal to us and our analysands regarding our relationship to a fragile world.

Thus, the book not only offers training candidates valuable support in their development process for our demanding profession, but also gives colleagues the opportunity to deepen their self-reflection.'

Dr. Angelica Löwe, Jungian analyst, Vienna, Austria;
member and lecturer at the C. G. Jung-Institut
Munich; editor-in-chief of the *Analytische Psychologie*

'In the theoretical part, Braun introduces key concepts of the Jungian theory such as archetype, self, complex, shadow, symbol, dream, alchemy, amplification, individuation, transference and countertransference and examines them from the standpoint of the encounter between the client and the analyst. His focus is the relationship. He looks at everything through this lens, and thus the concept of the transcendent function gains some depth. He demonstrates amplification in a way that gives the term new relevance and helps it emerge from its nebulous existence. Braun assumes that developmental psychology, attachment theory and infant research, as well as psychoanalytic object-relations theory, are all pertinent for the therapeutic relationship in the way Jung defines the *transcendent function*: the function that combines the contents of consciousness with the contents of the unconscious. This is what happens in the therapeutic space. Various terms have been used for this dynamic process, such as the concepts of the *interactive field*, the *intersubjective third,* the *analytical third, unconscious transference and countertransference*, the *analytical couple*, etc. In eleven chapters the author develops both the theory of this therapeutic relationship and the practice. For more than 50 pages he presents a case, which certainly contributes to the richness of the book. The details provide deep insight into the therapeutic process, enriched with dreams and drawings of the patient.

Braun uses the archetypal concepts in theory and practice as well as the intersubjective theory and practice, which together make this book so valuable.'

Dr. Isabelle Meier, Dr. phil., Dipl. Analytische
Psychologin International Seminar of Analytical
Psychology, Zurich, Switzerland

'Paying attention to the therapeutic relationship is one of the most reliable and clearly proven effect factors in current psychotherapy research. On the other hand, psychoanalysis has a tradition of more than a hundred and twenty years in carefully considering the complex encounter of therapist and patient, brought forward in terms of transference, countertransference, projection or intersubjectivity. Especially C.G. Jung's Analytical Psychology holds a treasure that should be used to improve our modern understanding of the bilateral and mutual development-path that therapist and patient walk together during their psychotherapeutic work.

For many years Claus Braun has worked on the subject of the therapeutic relationship, and as a well experienced psychiatrist and Jungian psychotherapy practitioner, his publications have always been a great win with psychotherapists. The present book summarizes his concepts and results and further develops them to a modern psychodynamic way of working in and with the therapeutic relationship. I am very pleased that the book, originally published in a German book series, has found its way to an international readership. I am sure that every psychotherapeutically active therapist will benefit greatly from it.'

Prof. Ralf T. Vogel, Dr. phil., Ingolstadt, Germany;
editor of *Analytische Psychologie C. G. Jungs in der
Psychotherapie*

To Katia

Contents

Foreword

I am delighted at the possibility of writing a foreword for this book. This book comes at a point in time and a psychotherapeutic landscape, from which one looks back almost in amazement at the time when members of various 'schools' vehemently argued about who was more successful, who had the better concepts, who belonged to the mainstream, who did not and who – precisely because he did not belong to it – may therefore perhaps be important. Meanwhile, we know from studies on psychotherapy that the general factors, such as therapeutic relationship design, combined with the expectation of improvement, such as the resources of patients, and the environment in which individuals live and in which they are treated, play a greater role than the various treatment techniques. In addition – and this is also shown by research (PAPs Study, 'Praxisstudie Ambulante Psychotherapie Schweiz') – many general intervention techniques are used today by therapists along with those specific to their schools, but more importantly many techniques also from schools other than those in which they are primarily trained.

Especially because we have found to have so much in common and readily take over unbiased intervention techniques from other schools, there is growing interest in what the 'other' concepts really are. As a Jungian, I keep noticing that Jung's theories are used as a 'quarry', whose fretted stones subsequently appear in a new design, or in a new 'setting', without reference to Jung. This happened with Jungian dream interpretation, from which many aspects are adopted wherever dreams are used today. That C. G. Jung was not the first to work intensively with imagination, but imagination is central to Jungian theory, has occasionally been 'forgotten'; schema theory certainly cannot conceal its proximity to Jungian complex theory, which arose 100 years earlier.

Much of this may happen because Jung's original concepts are known too little. That is why I welcome the publication of this book in which Jung's basic concepts – in their development – are described and formulated as they present themselves today, with a view to combining theory and practical work. I am sure that the Jungian theory with its great significance, the pictures and the pictoriality in it, can also give much inspiration to colleagues of other fields.

Verena Kast

Acknowledgements

In this book, I refer to my own psychotherapeutic experience in addition to the work of German- and English-speaking Jungian colleagues, who have for years worked on the development of the intersubjective perspective of analytical psychology.

I would especially like to thank Gustav Bovensiepen, Mario Jacoby, Jean Knox, Roman Lesmeister, Lilian Otscheret and Andrew Samuels for their content-based suggestions.

In the course of many years, my psychotherapeutic stance did not only expand through speech and silence in countless sessions with my analysands but was also enriched and deepened in conversation with Kurt Höhfeld, Renate Höhfeld, Wolfram Keller, Regine Lockot, Susanne Philipp and Anne Springer.

I thank Angelica Löwe for her valuable hints and the critical review of the manuscript.

Additionally, I would like to thank Ralf Vogel for his inspiring editorial support.

The careful and expert translation from German into English was done by Dr. Babette Gekeler, Berlin.

Chapter 1

Introduction

Understanding the therapeutic relationship has been of great importance in the psychotherapeutic treatment practice of Carl Gustav Jung's analytical psychology from the very beginning.

Jung had recognized at an early stage that while the analysand enters into a transference bond with the psychotherapist in the course of psychotherapeutic encounter, the unconscious psyche of the psychotherapist can also be stimulated, influenced and changed. He went as far as explaining the effect of induction of the transference projections possibly even causing a 'transfer of the disease to the person treating it' (Jung, 1946, GW 8, § 365). Arguably, the analysand and the analyst not only enter into a conscious relationship but engage in a relationship of mutual unconsciousness.

Communicative content in the space of the treatment room is therefore not only exchanged at the level of conscious encounter but unconscious sides of the psyche communicate actively, albeit subliminal, with each other. This latter form of communication may find expression in sympathies and antipathies, in reciprocal body language, in vegetative states, in moments of comprehension difficulties and in resistance, in the activation or weakening of defences, in moments of mutual encounter (Stern, 2005) as well as in feelings of abandonment. Furthermore, manifestations of such communication may be the experience of an 'unthought-of familiarity' (Bollas, 1997, p. 287ff.), the possibility of giving interpretations and accepting them, the shaping of dreams, and transference and in countertransference.

The energetic connection that is established between the analysand and the analyst in the course of the treatment might best be characterized using the concept of an *intersubjective field*, in which the *transformational energy* that the analysand needs for his development can unfold (McFarland Salomon, 2013).

While Jung mainly examined the contents of the unconscious and its symbolic forms of expression following the breakup from Freud, his early formulation of an interpersonal and intersubjective relationship model for treatment practice became neglected for a long time within analytical psychology. However, later research into early infancy articulated and described this model as fundamental a priori condition of any form of human communication. The examination of the intersubjective processes in analytical psychology has only recently regained theoretical and practical attention. Today, many Jungian psychotherapists pay particular attention to the circumstances of the early development of their analysands – thus working also with children, adolescents and young adults in order to improve their developmental outlook for the future.

Aided by the work of developmental psychology by analytical psychology-oriented researchers' findings of psychoanalytical object relationship theory, infant research, mentalization theory, and attachment theory could be joined into a conceptual body circumscribing the intrapsychic relationship dynamics that Jung described as transcendent function (Knox, 2011, p. 421ff.). The transcendent function describes the ability to connect contents of consciousness with contents of the unconscious. It is to be understood as a dynamic comparative process in which, on the one hand, explicit, conscious information is compared with the memories and engrams that are collected in our unconscious, inner working models as general relationship knowledge and that constitute the basis of our self-esteem. On the other hand, the transcendent function translates transformational energy into symbols, thus making it experiential and tangible.

Such process of filtering, adjusting and evaluating of real and symbolic experience underscores the meaning-making function in analytical psychology. This function in turn serves the self-regulation of the psyche containing the idea of a purposeful, functional and balancing extension of the conscious stance via contents of the unconscious.

Psychotherapy in the spirit of C. G. Jung sees itself more as a common process of reflection in dialogue and less as a dogmatic application of psychoanalytic explanatory knowledge. Jungian psychotherapy and psychoanalysis is crucially aware that behind and through the analysand's complaints rests a desire for integration and personality development or *individuation*, intentionally rooted in his own Self. A goal-driven process supports the use of the analysand's personal perspective enclosed in a particular cultural and historical context and aims at the development of the greatest possible freedom of thought and feeling.

The history of psychoanalysis can be described as a history of a change in the configuration of relationships in the course of the psychotherapeutic process.

In Freud's and his successors' classic standard technique, the analytical situation is conceived as a scientific investigation. Within rest only objects: the analyst in his 'mirror function' poses as object of transmission, the patient and his material as object of observation and interpretation.

In later Freudian object relationship theory, a two-person model of psychology, the analyst assumes the role of a real counterpart. The analyst makes himself visible as a human interlocutor, and the analysand can now experience new emotional relationships with the analyst, since he no longer hides in a mirroring function. The interpretation technique was supplemented by a relationship technique that allows for modifications of setting and forms of intervention.

Approximating the last 20 years, the paradigm of intersubjectivity has increasingly shaped our understanding of the psychotherapeutic process, which is now seen as a *fluctuating interactive mental field*. The analytic encounter is referenced by subjectivity and real relationships which offer stimulating spaced and potential for development beyond transference and countertransference. In this interactive mental field or matrix, common spiritual creations of the analytical pair are realized as an intersubjective *Third*. The intersubjective or analytical *Third* stands for the aforementioned new cognitive and emotional quality that the respective analytical pair produces uniquely. It need not be understood as something representational, but as a medium for psychotherapeutic processes of change and healing. Within such intersubjective field of psychoanalytic processes, both interactional-communicative competencies as well as the ability to gain 'insight' and connecting to one's own Self can grow and materialize.

Changes in the level of integration of the psychological structure based on such processes are also a prerequisite for better coping with intrapsychic and interpersonal conflicts. Conflict management competences and personality development intrinsically belong together and should develop through psychotherapeutic processes of gaining insight and configuring relationship patterns. Psychotherapeutic experience serves the *individuation* of the analysand and proposes a working model for his everyday life.

In this sense, a Jungian psychoanalysis is always directional in ways of being orientated towards the present and future while holding the conviction that perceiving and accepting the lines of development, which are deeply rooted in the personal psyche, form the basis of mental healing processes.

According to the experiences of my own psychotherapeutic practice and those of other colleagues (Otscheret & Braun, 2004), I concur that there is neither the standard patient nor the standard psychotherapeutic method. Each psychotherapeutic encounter is profoundly unique, taking not only an essentially individual course but creating and developing specific methods of healing grounded in the analysand-analyst's relationship. Whether this can happen depends primarily on the professional conduct of the analyst and his persistent effort to empathize with the analysand's life experience and relationship history. At the same time, he must be able to realize the persisting otherness of a stranger, which the analysand embodies, and who must nevertheless be warmly acknowledged.

This volume has two main parts. The first part (see Chapters 2–6) is devoted to the theoretical foundations of the therapeutic relationship from a Jungian perspective. Important theoretical differences to psychoanalysis based on Sigmund Freud are explicated. Subsequently I highlight the conceptual coordinates of Jungian theory formation, which are important for understanding the psychotherapeutic relationship and goals of an analytical process. Therein, an excursus on mind and brain will highlight the most important neuroscientific and developmental psychological research findings explanatory for the emergence of consciousness and interpersonal relatedness.

Subsequently, in Chapter 3, I formulate prerequisite personality traits and ethical requirements for both the fit of a person as analyst and considerations for the fit of the analytical matrix analysand-analyst.

Chapter 4 describes psychopathological concepts of analytical psychology, Chapter 5 the psychotherapeutic treatment objectives, and Chapter 6 the therapeutic space and the therapeutic framework and rules that govern the setting of treatment.

I dedicated the second part (Chapters 7–11) to psychotherapeutic treatment and relationship practice. On the basis of a detailed, anonymized treatment history, the initial phase, process and end of a Jungian analytical psychotherapy are described, focusing mainly on the therapeutic relationship and – so I hope – on accessibility to discussion (Körner, 2003).

I particularly focus on the reciprocity of unconscious influences on the workings and pathways of differentiation of symbolic contents stemming from the personal and collective unconscious. Moreover, I paid particular tribute to the dream work and the Jungian method of amplification.

The volume concludes by considering the special circumstance of German insurance-financed psychotherapy, its procedure and relationship to various dimensions of healing processes of the name of *individuation*.

I have chosen this particular treatment example in order to be able to present in a somewhat fine-grained way important changes and integrative steps in the context of the intersubjective dynamics of analysand-analyst matrix embedded in social relationships and the social reality of life. Although I have chosen analytical psychotherapy in a long-term treatment setting as an example, all essential aspects of content and psychodynamics can be transposed to other forms of psychoanalytic psychotherapy approaches with adults, such as psychodynamic psychotherapy and psychodynamic short-term psychotherapy. Various requirements for the analyst's conduct in the execution of analytical and psychodynamic psychotherapies are described in the 'Indication and dialogical processes' section of Chapter 8.

The peculiarities of the psychotherapeutic relationship in group psychotherapy cannot be described in the required detail in this volume because of its complexity. They are reserved for a separate work.

While in the text I use the masculine form throughout, all sexes are equally addressed, and I would ask you to consider and excuse any possible limitations of my text that arise from a gender perspective.

Passive connotations underscoring terms such as 'clients' or 'patients' are circumvented by the use of the 'analysand', of which I speak consistently notwithstanding the form of psychotherapy (whether psychoanalysis or other psychodynamic and short-term approaches) in order to emphasize the common task of examination and understanding in psychotherapeutic work. I use the terms 'psychotherapist', 'psychoanalyst' and 'analyst' synonymously for the same reason since the practitioner in any form of psychoanalytically informed practice tries conjointly with the analysand to explore the unconscious psychodynamics of his sufferings and make them accessible on the conscious level.

In some parts of the text, I use verbs such as 'should' or 'must', especially in relation to the analyst's activities. While such terminology ought not to give way to ascriptive demands for action or practical guidelines, I intend to stress what I might consider favourable or worth testing.

The bibliographic references should allow an extended reading of the respective terminology and the contextual lines of argument. The writings of C. G. Jung are quoted from the Collected Works in 20 volumes (GW 1–20). For reasons of comparability of different editions of the Collected Works and the English Collected Works, references include paragraph numbers. The quoted text passages are therefore made accessible not via the page numbers but via the year of the first publication, the respective GW volume, and the corresponding paragraphs (e.g. Jung, 1946, GW 8, § 365).

While the book is primarily targeted at the readership of psychotherapeutic colleagues and candidates for training and further education in all psychotherapeutic and psychoanalytical fields, I have tried to present all essential lines of argument in such a way that the text is understandable and accessible for other professional groups and interested laymen.

Chapter 2

The Jungian view of the psychotherapeutic encounter

From Sigmund Freud's psychoanalysis to Carl Gustav Jung's analytical psychology

The tragic development leading to the division of Sigmund Freud and his 'crown prince' Carl Gustav Jung, which affected both deeply, cannot be covered comprehensively here. However, at the heart of the debate that led to the separation of Freud and Jung lies a controversy about the essence of the mental conversion energy or libido. Jung did not want to see the libido restricted to sexual striving; he defined it as general psychic energy or life energy.

Both Freud's and Jung's varying models of the human mind were also subject to this controversy. To Freud, the unconscious mind was the place of suppression of incestuous desires and impulses that were too embarrassing and unacceptable to be held in consciousness. For Jung, the unconscious was '...the seed of eternal creative power, which makes use of old symbolic images but by which implies completely new spirit' (Jung, 1930a, GW 4, § 760).

Both psychoanalytic researchers had been personally known to each other since 1907, after Jung had sent his monograph 'On the Psychology of Dementia Praecox: An Experiment' (Jung, 1907, GW 3) to Freud after having been made aware of Freud's book 'Interpretation of Dreams' in 1900. He then applied Freud's theories about psychological trauma and its suppression in his investigations with the word association experiment (see 'The complex psyche' section of this chapter).

The two men got along right away. The intensive correspondence between the 50-year-old Freud and the 30-year-old Jung from 1906 to 1913 shows the extent to which their friendship had developed – a high degree of intimacy and familiarity (McGuire & Sauerlander, 1974).

At the first official meeting of the International Psychoanalytical Association in 1910, Jung was elected president and editor of the 'Yearbook for Psycho-analysis'. Jung was then re-elected as president three times until he resigned in 1913. After a long phase of retreat, introversion and self-analysis, Jung first used the term 'analytical psychology' in 1918 to demarcate his psychology from Freudian psychoanalysis. Since that time, the disciples and followers of Freud and Jung have often stood aloof from each other (Kirsch, 2007, p. 21f.).

In correspondence with their opposing views on the function of the unconscious and on the meaning of symbols, Freud and Jung differed above all in their method of interpreting products of the unconscious, such as dreams and fantasies (Plaut, 2004, p. 156ff.).

According to Freud, by eliminating repression and improving the recognition of reality, the analysand should become freer to enjoy his life more fully. Freud's demand on the analysand was merely that he has to accept some embarrassing truths about himself.

Jung also considered it extremely important that the analysand first of all should become aware of his personal contribution to the complaints of his suffering, his shadow. In the course of the analysis, the analysand should then also be made acquainted with the riches of his unconscious mind. In addition, cultural treasures of humanity should be revealed to him as a source of energy and creativity, which can lend meaning and value to his personal life.

Lesmeister (see Lesmeister, 2011, for the next two sections) describes the history of the Jung-Freud relationship as the story of a failed struggle for recognition. He quotes Jung, who writes in one of his last letters to Freud: 'I only suffer here and there from the merely human desire to be intellectually understood without being measured by the standard of neurosis' (McGuire & Sauerlander, 1976, p. 583).

The dialogue between Freudians and Jungians has not been reopened at any time in spite of numerous theoretical approximations. Thus, for example, Jung's concept of shared unconsciousness and unconscious communication mirrors ideas found in the models of projective identification and containment of the Kleinian school of psychoanalysis; the postulate of a shared, relational unconscious is today represented by the psychoanalytic intersubjectivists. Jung also began early to see the countertransference no longer as a disturbance of the analytic receptivity but to use it as a highly important body of knowledge consolidation (Jung, 1929, GW 16, § 163), as ensuingly formulated by Freudians Paula Heimann (1950) and Heinrich Racker (1959/1988).

Psychoanalytically worthy of discourse would be Jung's theory of autonomous complexes as part of a multifunctional model of personality and the total psyche. The complex model is well compatible with the concept of a network organization of brain activity without permanent centralized control, as represented today by brain research and neuropsychology (see Bovensiepen, 2019). The archetype theory, with which Jung aimed at the instinctive character of experiential readiness, finds empirical evidence in the results of infant research and comparative behavioural research, already recognizing infants as active and interactive partners and cooperative agents in social interaction.

Only when both sides can expect an overall gain, talking to each other and entering into a dialogue begins to make sense. Freudians and Jungians would have to become 'new-greedy' (German term *neu-gierig*, transl. *curious*) in wanting to earth out their different 'models of mind'. Among other things, we Jungian people would like to offer our idea that in addition to the limited individual psyche, there is also a shared mind that connects subjects with other subjects and even with matter. The existence of a historical collective consciousness like the *Zeitgeist* is just as indisputable as the existence of collective unconsciousness shared by large groups. The evidence is extensive and compelling, according to which the universal forms of the mind arise from interaction and communication with other spiritually gifted beings. What we know of the world seems to be more a matter of social construction than a reflection of facts of nature. References to the essentially relational and intersubjectively determined nature of the so-called *reality* increasingly come from non-psychological research directions, such as medicine, physics, biology and even the business world (Lloyd Mayer, 2002).

Oedipus complex and great mother

In a more profound sense, the separation between Jung and Freud suggests that they were both unable to see their key assumptions about the main human developmental tasks as complementary.

Freudians still place great emphasis on the importance of overcoming the Oedipus complex or the oedipal situation, the incestuous attachment to the opposite sex parent.

In contrast, Jungians believe Freud's understanding of the incestuous wish as too much explicit. The boy is not concerned with the concrete desire to unite with the mother or marry the mother. Rather, it is about a childlike desire to be further pampered by the mother, hence a yearning to remain unconscious, instead of becoming aware and take the hero's ride into the world. In a metaphorical sense, the fantasy of the mother incest also relates to a wish of being born again as a subject and individual in one's own right.

In terms of relationship dynamics in the Jungian term of 'objective level' the heroic metaphor relates to resisting the overpowering mother imago, the *Great Mother*, the mother-dragon. On the subjective level (or intrapsychic level) it relates to the task of the consciousness to fight against the power of the unconscious.

The archetypal role of the father is the embodiment of the world of moral commandments and prohibitions. He serves as a moral figure of orientation to show the son the way out of instinct and regression (Meier, 2015, p. 13ff.).

The Jungian Erich Neumann reads the Oedipus myth from the perspective of the hero's failure and a lack of consciousness. He sees Oedipus as a stuck half-hero who regresses to the stage of the son and suffers the victim's fate of the much beloved son.

Referencing a necessary development, the *individuation process*, Jungians see other mythologies as more fitting: for example, the Perseus myth and the birth of the creative, ascending Pegasus from the head of the dead Gorgon. Or the Osiris legend: the journey through the underworld, the night cruise through death and rebirth, which leads to the development of consciousness, to self-confidence. The tedious path of individuation and the contact with one's own creative possibilities and abilities must be fought anew by each individual (Löwe, 2014, p. 314ff.).

Excursus mind and brain: neuroscientific aspects of the therapeutic relationship

Today, the neural code seems to offer the most prominent approach to the question of how consciousness and unconscious, individual and collective elements relate to each other. In contrast, the genetic code (C. Levy-Strauss) and the archetypal code (C. G. Jung) imbued with scientific mysticism seem to lose importance (Bahner, 2002).

In line with leading cognitive neuroscientists (Wilkinson, 2010, p. 307ff.; Gallese, 2015), we can define man as a 'mind-brain-body being', resulting from the earliest and most fundamental experiences of relationships with the primary caregivers. The totality of all psychic processes cannot be separated from body experience. The earliest form of a spiritual organization seems to be the *image scheme*, the feeling of an embodied meaning (Johnson, 1987). Depictions of the *image scheme* are seen as an emergent spiritual form that develops out of physical experience and provides the basis for the ability to grasp abstract meanings for the physical world, the world of fantasy and metaphor – thus organizing experience (Knox, 2004, p. 69).

Experiences are stored in inner working models that provide the capacity for imagination and the development thinking, including the feeling for an inner psychic space and the feeling of one's own self. The evolving mind unfolds from the developing brain. New neural connections arise as a result of interactions with significant others throughout life.

Neural networks undergo critical periods during early postnatal development up to the second year of life, in which they are particularly susceptible to external influences. During these critical life periods neural structures are greatly adaptable (Roth & Strüber, 2014, p. 155ff.). Well into puberty the growing brain structures continue to be highly susceptible and vulnerable against environmental influences of psychosocial and material nature, such as drugs.

Cerebral cells mature under relapsing synapses at different locations in the nervous system. During such thrusts, far more connections are created than will ultimately be used. Environmental influences determine which compounds will activate and survive and which ones will not. Insufficient activated synapses are eliminated from the maturing structure (Solms & Turnbull, 2004, p. 234).

Conscious spiritual acts require the implicit presence of a spiritual subject, a Self. Whether this may be a primary Self is already created or exclusively of social origin seems to be difficult to decide. But it is certain that the Self is created and shaped through the learning and social history of a person. In particular, diverse and lifelong social mirroring processes are decisive, which help us to build a lasting spiritual self (Prinz, 2013, p. 71; Gallese, 2015, p. 101f.). Neurophysiologically, the so-called *mirror neurons* in certain areas of the brain seem to play an important role (Rizzolatti & Craighero, 2004), which provide background activity for imitation and empathy as well as for transference and countertransference operations.

Mirroring processes, as they are also the centre of interaction in the psychotherapeutic relationship, are also decisive for the restoration or stabilization of an unsettled, questioned, insecure spiritual Self.

The network structure of the neuronal interconnections of the brain, the brain-mind connectivity, allows the development of subjective experiences. One's own Self reflects the connection patterns of one's individual brain. Neurons organize themselves into groups and further into larger functional and dynamic clusters via emotionally synchronized and synchronizing amplifying circuits (Wilkinson, 2004). The memory systems and inner working models are to be understood similarly. These networks can be activated both interactively as well as independently or dissociated. It can therefore be said that these inner working models are emergent qualities or shapes of our memory systems. They are based on archetypical knowledge readiness. Their cohesion is created by connecting affects and emotions, which are set in motion basally via rhythmic imitations of the caregivers in proto-conversions (Molino, 2000, p. 175).

We find access to these networks through the increasing conscious experience. According to the consciousness model of Bernard Baars (2005), conscious contents appear analogous to spotlights on the workplace (*global workspace*) of inner working models. For example, if we direct our conscious attention to a human counterpart, then various unconscious brain regions are activated and the functions of a large number of specialized networks are coordinated, which previously worked autonomously.

The conscious experience is influenced by this: contents of our self-system, our intentions, our expectations, our perceptions, sensory stimuli, images and ideas interact with unconscious networks, such as memory systems, speech systems, automatic motor and stimulus-processing functions. Conscious recognition provides an activating and coordinating access to numerous parallel functions of the brain. These can be stimulated both by external as well as internal stimuli.

Inner speech activates our auditory consciousness. Visual awareness coordinates spatial memory and is involved in problem-solving processes. Macro-sensory experiences can visualize emotional and motivational processes, including feelings such as emotional pain, joy, hope, fear, sadness and so on. Motivational ideas are connected with parts of the motor cortex. Emotional centres, such as the amygdala, are also addressed through conscious content. Thus, numerous brain regions work together in order to ultimately allow inner goals and emotions to flow into plans and actions in the social context.

Conscious awareness is based on emotional sentience, in terms of background state of our consciousness. Emotion sensing is organized differently than the outward sensory modalities. This is mainly due to the fact that it is not a (sensory) channel-dependent, but a state-dependent function.

According to Panksepp (1998), four archetypically provided basic emotions underlie the universal affective reactions. They organize themselves into a search system, an anger system, a fear system and a panic system (with care-taking subsystem). The 'object side' of these systems is initially empty and open. Only through learning in relationships stimuli are connected to them, e.g. fear-anxiety reactions. These are often linked on the basis of a single triggering encounter. Such links can no longer be erased: nothing can undo the entry of the life-threatening object, the dangerous place or the disastrous situation contained in the list of dangerous things created in the fear system. In later life, only an output such as a fear or aggression response can only be deliberately inhibited.

A crucial feature of free will is precisely this ability to inhibit, i.e. the ability to decide not to do something. This possibility of postponing and evaluating decisions in the interest of reflection, of imaginary action, is organized in the prefrontal cortex (Solms & Turnbull, 2004, p. 128ff.).

Conscious recognition works with a double control/activation system: with voluntary attention on the one hand and with automatic attention on the other hand. The latter allows significant and meaningful unconscious stimuli to break through into the field of consciousness. Consequently, consciousness is always influenced and shaped by unconscious contexts. Dream activity and the complexes represent two unconscious forms of the differentiation of the implicit relationship memory.

An essential aspect of the therapeutic relationship is understanding what is going on both inside the human mind as well as at the physical-neuronal level during psychotherapy. Assessing the analysand's own ideas about the complexity of neuronal processes and what he thinks about anchoring knowledge and memory in neuronal networks are relevant information. In short: what the neuronal code means to him. Does the analysand have an idea of what could change at this level through talking, reflecting and experiencing emotionally the conflictual issues that emerge in therapeutic dialogue and through the symbols of the unconscious? On the basis of the neurophysiological investigations, one can know and if necessary show that at the level of the neural embodiment of mental restructuring basically only such contents become meaningful that can be experienced emotionally. The emotional marking of what is being exchanged on the verbal level is crucial for reshaping the synaptic interconnections and thus the network structure of the complexes. This means, for both partners in the psychotherapeutic process, that they have to get close enough that mutually emotional touches, 'impressions', can take place.

The complex psyche

C. G. Jung called his psychology complex psychology after the discovery of emotional complexes in his word association experiments. He considered a 'complex indicator' based on his experimental conclusions when extended length of reaction time to certain stimuli words indicated the existence of unconscious demeanour and attitudes that can influence the human ego consciousness – even dominate it like one part of the personality (Jung, 1934, GW 8, § 201ff.).

Jung defined the complexes as the living units of the unconscious psyche, as their actual building blocks. His *via regia*, or his royal path to the unconscious, is not primarily the dream, differing from Freud. According to Jung it is the complexes that cause dreams and symptoms (Jung, 1934, GW 8, § 210). The potential or energetic-libidinal power of complexes can be seen, for example, in neurosis and even more so in the neurotic dissociation of the personality. In certain conflict constellations, for example, a complex of fear can largely override the control possibilities of the ego-personality against its conscious will.

Our emotions and associations are organized through the emotional complexes: we feel and think according to our complexes. Complexes are organized around an archetypal core element and corresponding personal experiences.

The ego can also be regarded as such a complex forming the centre of our field of consciousness. It is responsible for our identity in space and time. Relatedly, important ego functions are memory, the willpower, the ability to self-control and impulse control, our relationship and attachment ability. We see the ego complex as the archetypal centre of consciousness, dynamically changing over time. The ego is the central element of the psychic structure in relation to the multiple facets of the Self, which could not be experienced without a conscious ego.

The ego complex is constantly influenced by energetically activated other complexes and may also be disturbed in its function. Examples are states of falling in love, grief, fantasies and fears. The resulting instantaneous or longer-term unconscious personality change is called complex identity. Complexes behave like autonomous partial psyches (Jung, 1934, GW 8, § 204); they can appear personified in dreams and psychoses. Because of frequent repression and projection of the complexes, we often cannot perceive them well enough.

The ego complex and other complexes can be linked into sudden complexes with long-term effects, such as religious conversion or the fear-anxiety reactions mentioned above. Complexes are also energetically charged via chronic affective responses, such as through an impaired stress processing or reward system (Roth & Strüber, 2014, p. 145ff.).

An active complex can put the ego into a state of lack of freedom, compulsive thinking and action. The effects of a complex become consciously experienced as impulses foreign to the self, like something happening to me from within, without my ability to control or influence it. At the onset of a neurosis, corresponding complexes establish themselves in the consciousness and penetrate the ego complex like an infection.

Different complexes often work together in a defensive organization against the self. In this respect, they influence the healthy and the pathological defence organization.

The Jungian perspective on the dissociability of the psyche varies fundamentally from the structural model of Freud and Klein: these assume a horizontal split (consciously-unconsciously); Jung's concept of dissociation rather means a vertical split into complexes or partial psychics, which contain both conscious and unconscious parts.

In summary, complexes create subnetworks, forming a matrix of all internalized interaction experiences (Bovensiepen, 2004, p. 38ff.). These influence inner working models, affects and expectation patterns, which are mainly stored in implicit memory and which are partly conscious but mostly unconscious.

Previously articulated complexes were grounded in a merely content-related understanding of complexes (Oedipus complex, mother complex, father complex etc.). Today, this understanding develops into a rather process-oriented dynamic model.

If the treatment process is heavily grounded in impaired and disruptive relationship patterns and the constitution of new relationship patterns, it might be useful to continuously refer to the parental complexes and their inherent mother and father imago in the working through of complex thematic tropes. Both imagines can have either a positively supportive or a negatively obstructive impact on the analysand. Moreover, they can be more or less nurturing growth-supporting and thus positively holding, or they can be restrictive, fixating and thus inhibitory of development. In addition, they can contain elements of giving and supporting inspiration and wisdom, and at the same time trigger impulses of disintegration, addiction, numbness and insanity (Dieckmann, 1991a, S. 14).

The scrutiny within inquiry about which effects of complexes are currently activated within the analysand as well as within the therapist and his countertransference should never wear off.

Function of dreaming

For Sigmund Freud, the manifest dream was a text encrypted by the dream censor, covering up the latent dream content. What counts as decisive is the latent dream content, which primarily contains repressed libidinal desires and childhood wishes. For Freud, the major task of dream analysis is to reverse the manifest, always compromising dream text back into the latent dream. Therefore, analysis concerned itself with the unconscious methods, such as condensing or displacement of this reversal or dream work has been performed by the patient. In doing so, Freud followed the ideas or associations pertaining to the individual parts of a dream. He asked to free oneself as much as possible from the impression of the manifest dream, because its symbolism is primarily governed by repression. The relationship context of the dream narrative, which is important for us today, interested Freud at best in the connection with the dream-triggering *rest of the day* and the *dreams from above* (Mertens, 2003, p. 50ff.).

According to the Jungian school, the dream is a spontaneous symbolic self-expression of the unconscious. The dream *is what it is* and therefore allows an undistorted view of the symbolization activity that integrates experience and expectation formed from the complexes of the unconscious psyche. The complex themes are presented in a dream, mostly *personified and dramatized* (Jung, 1934, GW 8, § 203).

Dreams are closely connected to our *Core Self* (see 'Self, identity and individuation' section of this chapter). They are scenic representations of ourselves, our desires and our fears as conveyed by the ego. Dream symbols of the Self (bowl, sun, star, gemstone, divine child, house and others) are often particularly fascinating.

Aspects of processing the experiential in dreaming serve the integration of experiences and the regulation of affects, especially in difficult and extraordinary life situations.

Complementary and *visionary functions* of dreaming presuppose a symbolic dimension of the dream life, namely a symbolizing, language-forming creativity of the unconscious psyche. Only if dream scenes represent the language of the unconscious can we assume a *communicative function* of the dream event that can serve to expand consciousness.

Therefore, persons, things and scenic processes appearing in the dream narrative should be understood beyond their real reference as the symbolic language of the unconscious, in which something initially unknown, even *foreign, is expressed within me* (Jung, 1954, GW 13, § 481).

The dream contains a symbolic message that wants to be understood by the dreamer, dressed often in metaphors. It serves the integration of experience and the preparation for life by supplementing or compensating the conscious attitude of the dreamer (Kast, 2006, p. 73ff.). Hence, functionally dreams adhere to mental processes that have not yet integrated emotional events of life.

Apart from playful, poetic and wish-driven dream activity of the brain, unprocessed information, especially unresolved conflicts and traumatic experiences, evoke the dream event as a strategy of coping and adaptation.

Inhering the character of a 'primary process', this is to say a prelinguistic thinking of the unconscious entails an uninterrupted activity of the psyche, which can be imagined as a permanent process of networking, alignment and integration between current sensory impressions and our conscious/ unconscious knowledge of experience. All memory systems – motor, sensory and emotional, whether procedural unconscious or declarative-conscious – are thus constantly modified and expanded.

Similar to learning processes, stable incorporation of new and prospective relationship experiences and expectations with affiliated affects into cortical long-term memory take place particularly during REM sleep, resulting in changes to our inner working models (Wilkinson, 2006, p. 300ff.). REM sleep, during which the most vivid and emotional dreams occur, carries a problem-solving function. This greatly facilitates information processing, thus enhancing performance in higher mammals (Mertens, 2003, p. 97ff.). REM sleep deprivation makes people susceptible to stress and increases impulsiveness.

A sub-symbolic, very rapid processing mode between different areas of associated cortexes may be of decisive importance for the emotional dream event (Mertens, 2002, p. 34ff.).

Dreams also serve to rework and shape our complexes when we perceive them as network structures in the matrix of all internalized experiences of interaction that were stored on the basis of their 'affective value' and the relationship fantasies associated with them (Braun, 2010a, p. 393f.). Neuro-physiological studies support the view that our 'psyche-brain' is constantly working and that dreams are constantly happening. The dream experienced and generated in REM phases is to be understood as an affectively marked sequence of the unconscious psychic processing of our experiences. Such a sequence can reach the field of consciousness when there is a certain degree of core consciousness activation (Solms & Turnbull, 2004, p. 224).

The picturesque scenic language of dreams is metaphorical and symbolic. The dreaming 'psyche-brain' uses vivid visual images or embodied simulations to process emotional states implicit in consciousness but not yet available (Wilkinson, 2006).

In summary, dreams have several intrapsychic functions:

- They are spontaneous current self-representations of unconscious activity.
- They are *mental processing modes* for not yet integrated emotional events of everyday conscious life, referred to by the remainders of the day appearing in the dream narrative.
- Dreams are used to *compensate or adjust* the conscious one-sidedness of inappropriate demeanours. Accordingly, nightmares can point to the workings of unconscious fear complexes.
- Dreams have a *future-oriented, prospective function* – an example would be the initial dreams in psychotherapeutic treatments, which often point out important aspects of the patient's development.

- Dreams can be examined in a *subjective* and an *objective* perspective. In the subjective view (*subject level*) all dream elements are related to the person and the psyche of the dreamer. The objective view (*object level*) refers to real relationships and life situations.

Considering cognitive scientific research, a suppressed drive can no longer be regarded as an energetic prerequisite of the dream. Rather the dream is the narrative-scenic mode of unfolding an inner micro-world in which the integration of experiences, the control of actions and the formation of identity happen as an emotion-regulating process (Moser & Zeppelin, 1996). The incorporation of experiences into the emotionally guided scenic long-term memory does not always succeed immediately. Certain scenes associated with either traumatic affect or failure cannot merely be integrated into memory as experiences. They remain affectively stimulated as non-integrated aspects of a complex and lead to the shaping of a dream as 'rest of the day'.

People organize their knowledge in narrative form. We assume that the functioning of the psyche consists of a multitude of parallel narrative processes. The psyche constantly attempts to make sense of experiences.

Dream images and dream scenes can also be understood as metaphors that expand into narratives and stories. Because metaphorical connections stimulate different sensory centres in the brain, such metaphors strengthen connectivity in the brain. Metaphors can be understood as connection patterns that stimulate the spirit of the individual and create a perspective, a contextual framework that gives explicit meaning to an implicit conflict dynamic appearing in dreams (Solms & Turnbull, 2004; Wilkinson, 2006).

Metaphorical narratives contain a certain perspective, an inner context that aims for understanding and refers to the personality of the narrator. Without an understanding of the content framework, which can be formulated as a structuring and guiding metaphor, we cannot know what a dream sequence means.

Through the symbolic metaphor events appear in a certain light and thus become an experience. The metaphor used does not simply document the event but rather creates a respective subjective perspective (Buchholz & Kleist, 1997). Dream metaphors refer to the as yet unintegrated parts of traumatic experiences, complexes and conflicts.

They are *intensity pathways* that affectively categorize experiences (Stern, 2005). As intrapsychic processing we can therefore see dream sequences as a test run of characteristic modulations of tensions in which the entire sequence produces emotional reactions and images, which are appraised as acceptable or not. If not, the dream makes a new attempt or aborts.

Aspects of dreaming that *process experiences* create the continuous revision of experiences and the reworking of events that cannot be stored, such as failure, trauma or special moments of happiness.

The *compensatory and prospective function* of dreaming presupposes a symbolic dimension of dreaming and a symbolic creativity of the unconscious psyche. This is the only way to assume a communication function of the dream event that can serve the expansion of consciousness.

Persons, things and scenic processes appearing in the dream narrative cannot be reduced to known circumstances and biographical real events. They are to be understood as symbols still to be developed, in which in the first instance something unknown is expressed. Jung (1954, GW 13, § 481) says about this form of confrontation with the unconscious that it is '...an experience of a special kind, namely the recognition of a foreign other in me, namely an objectively existing "different person", for whose presentation and communication the unconscious provides the symbolic form and the narrative framework'.

The analytical and psychotherapeutic situation and the interpretation of dreams encapsulate the concept of the individuation process as continuous de-integration and reintegration of self-parts (Fordham, 1985, p. 50ff.), since it is realized in the interaction of subjects. During this exchange and development process, the analyst in the beginning is provided with 'de-integrates' via dreams. Those elements could not be integrated up to now and their integration requires the inner work of the analyst (Springer, 2000, p. 121ff.). This way of understanding of the process includes an understanding of transference as externalization for the purpose of integration and counter-transference as a form of initially unconscious examination of externalized parts of the analysand's relationship experiences.

Typology: orientation functions and attitudinal types

Carl Gustav Jung began pondering the question what distinguishes his own personality from that of Sigmund Freud and Alfred Adler after realizing that their theoretical considerations were incompatible and could no longer be discussed. He based the issue of what might characterize the differences in their conceptions in thoughts around the problem of different psychological types. The result was his realization that every human judgement is limited by the type, and thus each view is relative (Jaffé, 1984, p. 211). Jung expressly did not intend to use his type description to develop a characterological method; it was merely to serve heuristic purposes (Jung, 1981, Letters I, Letter to Dr. Hans Schäffer, 27.10.1933, p. 171).

He defined four basic psychological functions: thinking, sensation, feeling and intuition. He distinguishes two rational and judging functions, thinking and sensation, from two irrational or perceiving functions, feeling and intuition. The two rational and the two irrational functions are contradictory and compete with each other. The most differentiated function is called the main function, one or two others are auxiliary functions, the least differentiated fourth function is called 'inferior', 'slow' or 'problem function'. Against the background of an extraverted or introverted attitude, all four functions can take effect, hence describing eight psychological types. Type patterns become further complicated by the fact that the conscious functions are confronted with having opposing unconscious groups of functions (Blomeyer, 1988, p. 101ff.).

Contingent on Jung's typology his theory of the Self and of individuation developed, based on the question of whether there is an inner entity that can balance or unite the multiplicity of these typological variants. The Chinese concept of *Tao* offered fertile ground for Jung's exploration of the Self (Jaffé, 1984, p. 211).

Typological reasoning and terminology are rarely used in analytical psychology today. The main reason for this is an inherent danger of misunderstanding terms such as 'main function', 'auxiliary function', 'problem function' and using or utilizing them in a fixed and rigorous evaluative matter (Rafalski, 2011, p. 173ff.). Another point of critique refers to the subjectivity of typological assertions. Different observers due to their own subjective typology will perceive specific elements of type in another person. In addition, they will receive reactions and responses according to specific typological aspects triggered by the observer or the situation (Blomeyer, 1998, S. 106). In other words: basic functions cannot easily be determined in therapy. However, an observation, whose application should be treated with caution, may be that auxiliary functions are used for parental purposes, such as caretaking, while primary functions serve first and foremost self-assertive purposes (Beebe, 2010, S. 74ff.).

The amplification of momentarily dominant functions of the analysand as well as of the analyst in their shared discourse is a useful tool in the course of an analytic process. Particularly, when the flow of understanding is ruptured or threatened. It can also provide helpful insides in the contrary case, when we suddenly feel particularly well understood or even emotionally carried by the analysand.

At times of difficulties during treatment one might profit from thinking about the inferior or problem function of the analysand as a potential influence in the making of such. Marie-Luise von Franz called it the 'personal devil, the personal inferiority of every individuum'. Sie adds: 'the little open door of the inferior function of every man adds to become the collective evil in the entire world' (Franz, 1980, S. 97).

Expending on the perceptive quality, namely intuition, can facilitate access to the *Analytic Third*.

The developmental school of analytical psychology contends the creation of an inner world based from the very beginning in the exchange processes between subjects that continuously and intersubjectively build relationship and share experiences. Already pre-partum, seemingly out of nowhere, a primary self is created, keeping ready the psychophysiological developmental potentials. These appear as archetypal expectations to the inner and outer world.

The formation of object relations materialized in a dual process of de-integration or bending outward these expectations as well as a reintegration or consuming of intersubjective experiences priory made. A major milestone in the development of the individual is the early ability to symbolize and to apprehend the *Third* in terms of ability to think and experience both – oneself and the other (Springer, 2000, S. 121ff.).

Archetypes and symbols of the collective unconscious

Withstanding the idea that our psyche is an empty psychic space at birth, Carl Gustav Jung assumed that the psyche is not tabula rasa but vested with instincts and the psychic readiness to react to our environment and primary caregivers, equating the life of animals. Jung called this this readiness to react 'achetypical'. He further called those patterns 'archetypes of the collective unconscious', to which all humans attend to their environment in a similar way. Archetypes are not visible per se but cause babies to instinctively search the breast or the eyes of the mother, reaches out for her when losing balance, searches someone to interact with and learn from. The readiness to create inner symbolic images representing essential and general life experiences, such as birth, relationships, marriage, motherhood, fatherhood, death and separation, seem archetypical as well. Archetypical patterns conglomerate in a singular personality which thrives to create an identity mirroring an unmistakable version of the endless possibilities for an individual to *be*. Apart from their instinctual nature they seem to also contain dynamic qualities of a self-structuring complex, out of which new archetypical pattern can arise (Saunders & Skar, 2001).

Archetypical images appeal to us emotionally since the abstract but libidinous-energetically charged archetype unfolds pictorially and symbolically fascinatingly within them. Myths and fairy tales, as they were first passed down orally, later in writing over many generations, can be understood as staging of archetypal meanings. The imaginary worlds of the gods, the astrological systems, all possible variants of the faith in fate can be interpreted as metaphors of archetypal ideas about the nature of one's own psyche and the social world. There is much to suggest that archetypes also have their own formational history through the historical expansion of human consciousness. An enormous resonance can be found in the vivid mythologies of many attractive Internet games.

In sum, archetypes are initially impersonal and innate structural schemata. They bind together with unconscious personal experience and create archetypal images and archetypal nuclei or attractors of complexes, which help organizing personal experience and the symbolic world of the human psyche. The symbolic world becomes accessible to us though dreams, or imaginative or creative impulses of various sorts, and according to the availabilities of the therapeutic pair.

Three thematic circuits of the collective or shared unconscious seem to be most accessible for the personal conscious experiential realm: Jung called them the *shadow, animus/anima and the sage* (Jung, 1933). The archetypical area of the *shadow* (see section 'Shadow integration' of Chapter 5) contains all personality attributes I reject in myself, judge or don't want to accept. *Animus/anima* (Jung, 1933, §§ 55ff.) figuratively indicate all that fascinates, inspires, rejoices or seduces us. They are frequently experienced as projections onto the opposite sex or like a compulsive fixedness and remain highly ambivalent in their outcome. The *sage* appears as potential figures of council in dreams and imaginations (see section 'The integration of creative expressions' of Chapter 10, Fig. 10.4). Capturing and interpreting the symbolic content of the metaphorical language and the scenic Gestalt of dreams as a natural language of the unconscious requires a lot of reflection and is facilitated via amplification (see below).

The unconscious forges symbols as carriers of energetic quality for libido, able to be guided towards cultural or intellectual interests. Jung called the symbol a "psychological machine, that transforms energy" (Jung, 1928, GW 8, § 88). The discovery of magically appealing symbols has been characteristic for human societies from their very beginning. These became religious, meaning cherished in a spiritual connection from which the group and the individual could draw psychological strength and orientation.

The human ability to symbolize grows out of the biographical experiences of shared attentiveness between child and caregiver predating language. Further, the child playfully learns that everything can be understood as something else: 'This is a thumb, that shakes the plum, that one here picks it up, that one there brings it home and the little one here: eats them all up!' (German finger game comparable to the sing-a-song: 'One, two, three four five, once I caught a fish alive, six, seven eight nine ten, then I let it go again – why did you let it go? Because it bit my finger so. Which finger did it bite? This little finger on the right'.) This means: all things symbolic are validated interactively. Symbols are therefore to always be understood as intersubjective and arising from a certain perspective. We construct our reality and it's meaning through the aid of symbolic manifestation. Human onset of consciousness happens on the basis of an inner meaning-making and significant content mediated by symbols. The universal language of symbols follows an associative, emotionally laden and intense logic. It binds together conscious, preconscious and unconscious (Dorst, 2015, S. 20ff.).

An essential target of Jungian psychotherapy is the development of a therapeutic relationship in such a way that a shared gazing onto the inner world of symbols and images becomes a naturally given part of the therapeutic dialogue.

Alchemy, transformational processes and psychotherapy

Jung's empirical rediscovery of the *objective psyche*, meaning a commonly shared (*collective* unconscious, carried immense meaning (see Franz, 1977).

The mental layer of the commonly shared unconscious depicts a level of structural identity, which all people of at least one cultural framework share. The commonly shared unconscious is a dynamic and continuously changing process of becoming, which dynamic is due to the human spirit and it's inventive, inspiring and imaginatively symbolizing activity at any particular time and in interrelation with archetypical images (Franz, 1978, S. 153).

Furthermore, the commonly shared unconscious does not only exist in the individual psyche of a person but this person is surrounded by it in a wider psychic space. This idea is frequently shared by religions believing in the wandering of souls, the resurrection of the dead or the producing of a fine material resurrected body like the Egyptian priests, conducting mummification to transubstantiating the dead to Orisis.

The 'spiritual' field of alchemy (in contrast to chemical alchemy) was initially concerned with the creation of an immortal body captured in Gnostic belief of Anthropos or a divine primal human having fallen into matter and who was ought to be freed from his ordainment to the passing material world through the alchemical opus. Consequently, alchemists suspected a soulfulness or an unconscious in matter.

Within intellectual history alchemy posed as compensatory undercurrent to a male-dominated dogma of Christianity, to which it related in a similar way as dreams would to consciousness (Jung, 1944, GW 12, § 26).

C. G. Jung showed how alchemists would project their own unconscious into something material through ritualized processes (Jung, 1943, GW 12, §§ 332ff.). Essential to the processes were the production of eggs, primal waters, philosophical gold, the *Lapis philosophorum*, the panacea and so on. Those essences appropriate the collective unconscious contained and effective as well within each self and process of individuation.

Alchemists describe the following phases of chemical metamorphosis (fermentations) via the colour change in their alembic and characterized them as blackening (*nigredo*), whitening (*albedo*), yellowing (*citrinitas*) or reddening (*rubedo*). Apart from the initial substance, the *materia prima*, water (*aqua permanens*), fire (*ignis noster*) and the smelt oven representing the carefully enclosing and protecting miracle pot (*vas mirabile, vas hermeticum*) played a crucial practical and symbolic role.

Both the relationship to astrology (planets also being gods, seven metals of the alchemists representing planets: Mars is iron, Saturn is lead and so on) and the mental moods were together considered as paramount for the success of the alchemistic endeavour. Therefore, personal inner states became influential to the outcome of chemical processes. The male alchemist finds support in a female counterpart. His support stems from a female figure of transformation (*anima mercurii*), with whom he achieves the union of opposites (conjunctio) in the mystical marriage of the alchemist oeuvre.

With a view to the unconscious psychological claim of the alchemists, it becomes possible to relate and amplify elements and symbols of alchemist processes of transformation to psychotherapeutic developmental phases and processes of transference.

Referring to the psychotherapeutic developmental processes, the alchemist blackening or nigredo parallels the *dark* initial stage of psychotherapy (Jung, 1945, GW 16, §§ 376ff.) insofar as early transference processes, such as shadow projections, and an unconscious analysand-analyst identity form and perform, which significant *initial dreams* can indicate.

Unconscious content is the psychological equivalent of the *prima materia*, whose strenuous contestation is the primary work of the analytical process. Experiencing a reddening/rubedo, a psychological 're-vitalizing' is a capturing experience for both analysand and analyst, whose surfacing is always felt as somewhat mysterious.

Estrangements and uncertainties are typical for both parties when dealing with one's own projection and unconscious and intensive compassion is asked of the analyst when accompanying the analysand on *his way to himself* and a meaningful life (Jung, 1945, GW 16, § 400).

Jung used the fascinating image series of an alchemist text, the *Rosarium Philosophorum*, to augment the presentation of transference phenomena (Jung, 1945, GW 16, §§ 402 ff.; see also Wiener, 2017, p. 78 ff.).

Fig. 2.1 (on page 27) describes the complex psychodynamics of transference processes. The analytic goal is the symbolic renascence of the analysand.

Self, identity and individuation

Analytical psychology is particularly interested in a person's psychological development and process of finding identity, his or her individuation. The process of individuation is unique in every human being. It is understood as a process of getting close to oneself, in which the ego gets an increasingly conscious contact to one's own self and thus to the creative potential of one's own unconscious.

Individuation is an internal, subjective and integrative process which simultaneously happens in constant relation to the other (Jung, 1945, GW 16, § 448). Research findings from developmental psychology of modern infant research and neurophysiological research findings on the development of neural networks of the brain both show the intersubjective nature in the genesis of self.

The self learns to observe itself from the outside and to take the perspective of the other through interaction. Consequently, it acquires its consciousness through self-reflective mirroring experiences with others (Altmeyer, 2000, S. 206). In particular, neurobiology assumes the ability to view oneself through the eyes of the other as crucial intersubjective antecedents for the development of a feeling of identity.

At the heart of a positive identity development rests the intersubjective human appreciative recognition (Benjamin, 1996). Paradoxically, aggressive impulses play a major part in relating to the other not just intrasubjectively but in relation to the other. As the term aggression (lat. ad-gredi) connotes a closing in on something, aggressive impulses determine a reaction objectively placing the other outside of one's own self. Winnicott describes such dramatic attempts of babies in the destruction of the object as resulting in the survival of the other, who then lets the infant know that he or she is not merely a product of its fantasy (Winnicott, 2002, S. 133f.).

Within the developmental process we experience a fundamental tension between the denial and acceptance of the other, between omnipotent fantasy and acceptance of reality. By accepting the parting mother as not being bad, but independent, the child gains its own independence. Ideally it can reach a self, which intra-psychically can allow different voices, asymmetries and contradiction to exist, and can withstand ambivalence and renounce the desire to have a perfectly uniform consciousness.

Individuation as principally organizing process of experience happens inter-subjectively in a living system, which is characterized by an overlapping and interconnected play between different subjective worldviews. The traditional idea in psychoanalysis that a successful constitution of self in childhood is sufficient to meet later requirements for managing life is now criticized as deficit of an individualistic theory (Lesmeister, 2009, p. 255ff.). Psychotherapeutic processes today need to pay attention to identity-relevant factors in the dynamic changes throughout the life course. Individuals need to adapt to those factors but equally need to challenge their individuation in opposition to those.

From the start we are both individual creatures and creatures of relation. The need for conversation and dialogue is genetically-archetypically anchored. The neurophysiological heart of our social identity, the orbitofrontal cortex, need conversation to develop. The mother's rhythmically-melodic chanting modulates first narrative cycles. Further, resulting emotions lead to morphogenetic regulation in time differentiating the primary sense of identity. Thus, a dynamic-emotional syntax emerges, i.e. an emotionally coherent pattern system or set of rules for the use of language (Trevarthen, 2010). Moreover, a rhythmic-affective semantics surfaces connoting an emotionally rhythmized attribution of meaning (Molino, 2000). An epigenetic deep grammar develops. The ability to symbolise and use symbols (speaking) unfolds intersubjectively through continuous mimesis or imitative play. The experience of a self in the relationship to others is the central organisational principle of spiritual and meaning making structuring (Mizen, 2009, p. 269).

Subjective self-perception is present long before language acquisition. The comparative perception of the infant is structured by *amodal* perceptual processes that go beyond the mere sensory impression. They comprise of transmodal communications, the seeing of faces and bodies and sensing one's own bodily existence. Self-esteem arises once the organism can construct an image of itself and a perception of how it is influenced by others (Mizen, 2009, S. 268). Resulting is the experience of an emerging self and an emerging relatedness (as from the 2nd or 3rd month of age). As from the 7th month the development of a *Core Self* (*Kernselbst*) or *core-relatedness* can be observed in children. Children then become aware of a feeling of authorship, self-coherence, private affects and their own history. The subjective self is now established based on the working model of intersubjective experiences. Therein, affects contour the combination of satisfaction, excitement and motivation at any point in time.

From the 15th month onward the *verbal self*, comprising the ability to act symbolically, emerges. Simultaneously, the ability to social categorisation of oneself and others surfaces (Stern et al., 1992).

Unconscious self-esteem and the feeling of self-efficacy arise through the internalization of a relationship dynamic of self and others through experiences in the interplay of language and answer (*turn-taking*) and the coordination of moods with the primary caregivers. Rather than being a uniform structure, the self is dialogical or diverse, consisting of a cluster of experiences of the *self-with-other*. Which aspect of our self becomes effective at any given time remains context dependent. Meanwhile, self-confidence remains somewhat unstable, since not only secure attachment experiences, but also traumatic memory systems remain effective for life (Knox, 2012).

Attaining *I*-identity means accepting the expectations of others as well as being able to take distance from them and expressing one's unique individuality through language, creative output or other behaviour. *I*-identity is part of the interaction process itself and remains in constant need to be renegotiated in light of the expectations of others and one's own changing life course. Contradictory expectations and norms require a person's ability to influence them critically and creatively or to superscript them (Krappmann, 2010, S. 208–209).

Potentially, every person is a *Homo creator* (Neumann, 1957). While the relationship between person and society is often conflictual and tragical, it is also essentially creative, since the archetypical is creative and the Self has the capacity to use the archetypical or its images without being bound to it in obligation. This is the "divine" character of the Self. It stands in paradoxical union with the Self of all other individuals and this connection becomes manifest and experiential for example in transference and the *participation mystique* (Jung, 1945, § 376; see Chapter 2.2).

According to the well known Jungian Erich Neumann, the fundamental constant to human existence is a strict belief in the indestructability of the connection between *Ego* and Self, hence the indestructible link to creative, consciousness-evoking and divine potency of the archetypes. Ways in which life is being carried out in reality is embedded in the collective consciousness, the *Zeitgeist*.

In earlier times, the willingness to accept prefabricated identity packages was the central criterion for coping with life (Erikson, 1988). Today it depends on individual identity work, the ability to organize oneself, to become active or embed oneself in one's Self. The success of this identity work is expressed internally through feeling coherence and self-recognition, authenticity and meaningfulness. To the outside world success manifests through recognition of our self-portrayals in relationships, in our work, in the acceptance of our value orientations and our core biographical narratives (Keupp, 2012, pp. 96–97).

Our state of maturity today is less reflected in our ability to control needs and the environment, that is to say in our overall strength of self, but in our ability to be open towards the manifold animating elements of others and our own person (Honneth, 2002). In postmodernism, our personal identity develops more as a diversity of identities able to remain 'capable of plurality' or 'transient' (Welsch, 2002, p. 31ff.). Living in today's world is marked by the requirement of acquiring 'identity competence', namely the ability to form context-specific new facets of the self and to continually redesign them. Ongoing work on identity should make it possible to repeatedly coordinate and integrate individual partial identities anew (Döring, 2003, p. 326ff.).

One danger rests in the possibility of getting lost in the formation of one's own *persona*, meaning the perfectionism with regard to the masks and societal roles in which we operate in our social environment: '...to free one's self of the false covers of the persona on the one hand and the suggestive power of unconscious images on the other' (Jung, 1920, GW 7, § 269).

The meaning of individuation contends the need to being open of the analysands' new, unexpected possibilities of being, in which they are emerged (Lesmeister, 2009, S. 20ff.). Regarding the therapeutic relationship, this means constantly being made aware that the analysand is encapsulated by an unpredictable process of becoming, and whose standards might be very different from our own.

Psychic life as symbolic stage

It seems to be an expression of personal individuality to what extent the analyst is inclined to approach the reality of the soul more intellectually or more imaginatively. Wolfgang Giegerich (2012), a Jungian thinker, points out that there can be no psychology worthy of its name without soul. He emphasizes the reality of the soul in modern times. There is a need to differentiate simulated soul phenomena, such as certain forms of spirituality or transcendental belief from the logical life of the soul. While, according to Giegerich, the soul appears today in the neurosis, it is yet confronted to the need of overcoming it at the same time through human consciousness.

For many Jungian psychoanalysts, the language of imagination, fantasy and images is still as important today as rational or intellectual concept formation. The pictorial Jungian figurations of the psyche try to come as close as possible to the emotionally rich, lively and occasionally dramatically escalating psychological events in a way that at first resembles more the attempt of a holistic-emotional perception than a rational-anatomical dissection of mental events and experiences. Many figurative images of conscious and unconscious experience that C. G. Jung has shaped are so obvious that they have long become common knowledge, such as the term 'archetype' or 'archetypal', the Old Wise Man or Woman or Anima and Animus.

In the Jungian view of the internally staged soul life, there is the more or less well-illuminated area of ego-consciousness, outside which the large dark area of the personal and collective unconscious extends, circumscribing the sphere of the comprehensive self. The self emerges from the diffuse but expectant matrix of the psyche through the internalized experiences of relationships. At the same time, the intersubjective experiences form sub-personalities or complexes, among which the superficial yet main role is played by the ego complex, i.e. everything that constitutes me in my self-perception, my perception of others, my self-control, my possibilities of communication, my ability to relate, my will and ability. The conscious side of the ego complex is more or less under the influence of its unconscious *hinterland*: the unconscious feminine, the anima or the unconscious masculine, the animus. They are aspects of the mother complex and the father complex, into which all experiences with women and men, including our ideas about our own ancestors, collective-archetypical images and cultural complexes, are incorporated.

By means of contrasting pairs such as conscious-unconscious, female-male, Anima-Animus, father complex – mother complex etc. –a peculiarity of Jungian dialectical thinking becomes clear: psychological development is seen as an expression of overcoming or integrating opposites or imbalances. For example, a conscious and unduly one-sided orientation towards prestige and success requires equilibrium not to lapse into a superiority complex on the one hand, not an inferiority complex on the flip side. Such equilibrium can be initiated both through a conscious confrontation and through the balancing influences of the unconscious through dream symbols or other creative-symbolic impulses.

In concrete everyday life encounter, those complexes are activated by backdrop experience that seems to fit the respective situation. They unconsciously and instinctively constellate our emotional willingness to react and our available possibilities for action. On an emotional level, a fear complex, an inferiority complex or a fantasy of superiority can become so strong that such a complex temporarily takes over the complete guidance of my feelings and actions.

In the later treatment example, one-sided pathological aspects of the parent complexes played a central role in the latent guiding of fears and suffering of the analysand. Through his unconscious anima complex, combined with his creativity and his longing for relationships, he was repeatedly drawn away from the parental imagines onto the path of independence and capacity for love. However, this aspiration could only succeed when he had found access to the inner image of his autonomous masculinity.

Transference and countertransference in analytical psychology

The traditional psychoanalytic concept of transference assumes a distortion of the analysand's perception, which becomes apparent through shifts and projections of biographical conflictual contents onto the analyst's person. This point of view calls the analyst to be only interpretive in interaction while assuming the role of a pure observer. This role makes him available as a mirror or projection screen for the analysand's fantasies.

C. G. Jung's understanding of transference confers both analysts and patients' conscious and unconscious aspirations to an intense engagement as persons and to intertwine projectively. He characterized the transference as Conjunctio or Hierosgamos/Holy Wedding (Jung, 1945, GW 16, § 458) and compared it with a chemical reaction of two substances. Both must change in order for what can be understood as a *new connection* to establish. Jung understood the therapeutic process as a content-related and energetic exchange between the consciousness and the unconscious of both participants, in which unconscious contents are mutually projected or transferred and compared (Figure 2.1).

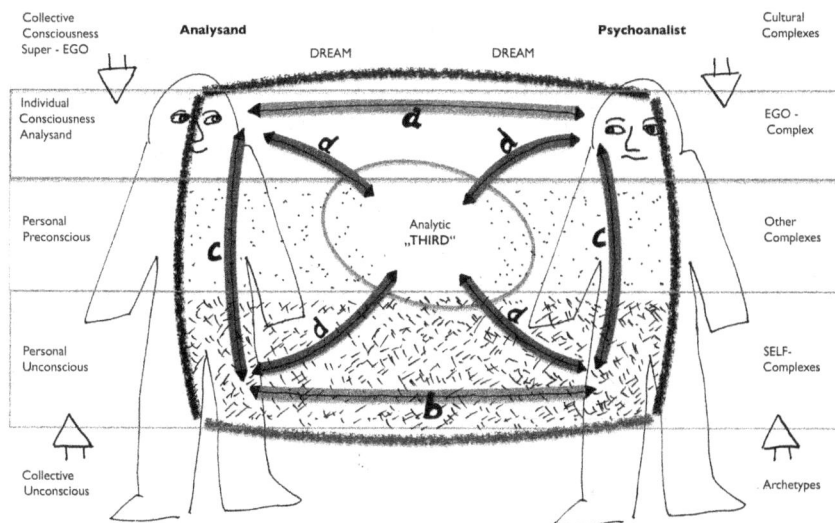

Figure 2.1 The analytical space/the transference matrix (cf. Jung, 1945, GW 16, § 422, 'Heiratsquaternio'; Wiener, 2017, p. 80, p. 95ff.).

The direction of the arrows describes mutual influence between analysand and analyst (a), between both unconscious (b), between both conscious and unconscious psyche (c), and between the conscious side of one and the unconscious side of the other (d). Small dots: level of the personal unconscious. Dotted: level of the shared ('collective') unconscious.

a – describes the conscious mutual relationship.
b – remains largely inaccessible but is emotionally noticeable in sympathies and antipathies.
c – describes the degree of personal entanglement of both in complexes and neurotic attitudes.
d – is the psychoanalytically fruitful level of relationship from which the *Analytic Third* (see below) can unfold.

Awareness relates to the withdrawal of my projections or transferences, i.e. the conscious perception of my projection or transference activities. C. G. Jung depicts the extent to which transference consist of the projection of ideas that originate from archetypal patterns and to what extent they result from the actual life experiences of the individual. Despite his focus on the archetypal aspects of transference, he was one of the first psychoanalysts to recognize the clinical importance of countertransference. Contents of the countertransference originate on the one hand from the unconscious episodic relationship memory of the analyst; on the other hand, they can also be influenced by his own neurotic demeanour and residual neurosis, which may protrude despite solid analytical self-awareness.

The metaphor of a psychic infection can serve to illustrate the neurotic countertransference: the analyst identifies particularly easily with his analysand when both unconscious motives and dispositions are similar (Jung, 1945, GW 16, § 365). Complex constellations can also lead to a disturbance of the ability for inner dialogue in the countertransference (Jung, 1934, GW 8, § 199ff.). To such, we analysts ought to always pay attention and which should also be the subject and task of our own supervision or intervision.

Decisive for external objectivity and the validation of the analysand's arguments is therefore our ability to engage in inner dialogue and to deal with the voice of our own unconscious. Those who cannot accept the other cannot concede the right of existence to the other in themselves and vice versa (Jung, 1934, GW 8, § 187).

The Analytic Third *and the creation of analyst-analysand pair*

Sigmund Freud wrote in a letter to Carl Gustav Jung on 6 December 1906 that healing in psychoanalysis is achieved by transferring the intrapsychic unconscious libido to the therapist, who makes the energy available for the translation of the unconscious. And he added: 'Healing is achieved through love' (McGuire & Sauerländer, 1974, p. 13).

Transference organizes the psychoanalytic process to a high degree by way of affects and moods. The process becomes effective through them and through the staged symbolic representation of the relationship to mental images of parents and other important caregivers.

The transference of the analysand corresponds to the transference and the countertransference of the analyst. In their interaction, both parties react subtly to each other and actively shape the therapeutic relationship. The analyst is involved and thus, even if he wanted to, he could not escape his complicity and co-responsibility through passivity, such as silence, idleness, restraint, or with reference to his abstinence. Consequently, neutrality and abstinence have just as intensive an effect on the analysand as active actions (Bettighofer, 2016).

Transference and countertransference are intertwined intersubjectively. The analytical process changes both the analyst and the analysand. In the common intersubjective field, those emotionally intense spiritual touches occur between the two. Jung calls this 'participation mystique' and Daniel Stern calls them 'Now Moments or Moments of Meeting' (Stern et al., 1998). It is there where development takes place through affectively significant new experiences of relationships. These new relationship experiences are conveyed by the intersubjective *Analytic Third*. The *Analytic Third* is created as a result of the process of change that the analysand and the analyst enter into through their encounter. In form of a co-constructed figure-ground model, both can be understood to emerge represented as analysts and analysands. The term *Analytic Third* is used to describe the qualitative change that turns two people into a representation of a couple and a child, thus mother and father into a family. It is subjectively experienced differently by analyst and analysand, yet the experience of the analysand remains at the centre of the analytical dialogue (Ogden, 2001, p. 23).

In the experience of the *Analytic Third*, the analysand may come to feel a healing relationship with another human being for the first time in his life. This can be an experience calming, supportive and symbolically reflective. Intrapsychically, such intersubjective experiences promote the development of a positive Ego-Self relationship and thus a positive self-esteem.

Transference and the interpersonal relationship

The heterosexual couple ensures the survival of mankind through their sexuality. The erotic desire of another in the sense of the Oedipus complex in boys, the Electra complex in girls or in the sense of a primary homosexuality shapes personal and sexual development. Erotic desire is therefore an essential source of individuation. Oedipal constellations (child, mother, father) are an enduring characteristic of human relationships, because people develop in triangular relationships – also in psychotherapeutic space. Erotic desire is translated into imaginary of sexual fantasy, wherein patterns of relationships with the early childhood caregivers re-emerge. If it becomes possible to talk in the treatment room about dealing with pleasure and satisfaction in sexual experience, a reflection of these patterns can also be found in transference. Here, forms of interaction of the analysand become accessible, such as for example dominance and submission, advertising for oneself or rejection of others (Quindeau, 2014, p. 103f.).

The preconscious basis of our sexual life rests in the image of the sexual connection of our inner parents, with whom we have identified ourselves introjectively in both the male and female roles (Meltzer, 2007, p. 98ff.). An image of positively united inner parents constitutes an inner attachment to responsibility and thus to the capacity for non-incestuous parental love. The term 'aimance', stemming from Françoise Dolto, denotes this love and contrasts it to the sexually connoted 'amour' (Dolto, 1996, p. 419ff.).

The analytical framework, the abstinence and the asymmetry of the relationship prevent erotic fantasies for being exerted, even if they become aware. This frustration can cause the analytic process to become cold or seemingly unfruitful.

Metaphorically, the analytical pair then lies together like waxwork until an *Analytical Third* appears (often via an alive dream symbol), making visible what has thus far been bound in averted sexual images. This includes relationship and individuation fantasies and contained longing for transcendence, for exceeding one's own limitations (Guggenbühl-Craig, 1976).

In order to modify or dethrone the figure of transference, the experience of a non-incestuous relationship with the analyst must become possible, allowing for separation and divorcement. Figuratively speaking, both the analysand and the analyst have to trespass from 'object to *the other*'. Decisive here is the ability to perceive the other person in his independence and dignity. In this way the other person can be experienced as such and beyond an *object* of drive or desire in one's own perception (Küchenhoff, 1999, p. 200 ff).

Overcoming the analyst's own incestuous impulses in the mutual process of de-objectification is paramount. It requires the analyst to intensively examine possible resistance to eroticized countertransference.

In the *Remarks on Transference Love* (1915, GW 10, p. 313), Freud speaks of the necessity of suppressing countertransference. Adjacent, Jung points to the inevitability of countertransference despite any great effort involved in clarifying the unconscious, mutual, often chaotic seizure of affects in transference and countertransference (Jung, 1945, GW 16, § 383). Today, professional expectation requires the analyst to become as aware as possible of his countertransference. One problem is that becoming aware of the countertransference always takes place retrospectively through perceptions of one's own unconscious enactments. These can be vocal changes, changes of body position, somatic events or *actualization* of one's own unconscious desires (Cambray, 2001, p. 277ff.). If such a desire in countertransference would be manifested in an oedipal-erotic desire towards the analysand, the analyst task should be to change it into a therapeutic stance of trans-oedipal *aimance*: a love beyond sexual desire similar to one based on taking responsibility for a *child* and based on language and understanding (Meltzer, 2007, p. 184). The clarification and overcoming of one's own adolescent-sexual impulses facilitates an inner demeanour of parental eros and non-incestuous *aimance*.

Based on the above, moments of intimate closeness can grow, through which analysands can relax, open up and confide. Both, analyst and analysand, are in connection with their own Self and in good contact with each other. In these moments of cognitive encounter, fundamental feelings of love for one's own Self and for the other create the energy of transformation that enables mental restructuring.

Another *inner work* of the analyst is the reflective triangulation of the analysand's individuation process (Klingenburg-Vogel, 2008, p. 194ff.).

His *external work* consists of encouraging self-reflection and developing a stance that recognizes the analysand's dependence on the 'other' in the intersubjective sense, enabling the development of creativity and life worth living (Winnicott, 1992, p. 78ff.).

Affective realizations must be accompanied by their linguistic symbolization. An sophisticated inner image of the feeling of being loved needs language! In an interplay of reverberation and linguistic communication with *aimance* and acceptance, an affectively valuable, transformative *Analytic Third* arises, characterized by a lovingly unique intersubjective dialogue.

The development of *aimance* and acceptance paves the way to reaching the end of the common analytical path marked by a separation of letting-oneself-go in a non-frustrating farewell from the analytical situation. However, sometimes, as in our treatment example, pain of separation and de-idealizing the analyst spawns an aggressive potential necessary for the analysand to go his own way.

The unique encounter of subjects: intersubjectivity, reciprocity and independence of the intrapsychic

Classical interpretation technique is grounded in neutrality and abstinence. As mentioned above, classic analytical stance was based on the ideal of the analyst's anonymity and the illusionary assumption that the analytical arrangement could be kept free of relationships.

Within psychoanalytic advancements, namely object relationship theory, and its emphasized view of the internalized object relationships and their re-staging in the current transfer (two-person theory), the personal relationship to the analysand becomes thematized. The *Leitmotif* of a possible corrective emotional experience shifts the relationship into the centre of the new technique, according to which the I is not primarily looking for drive satisfaction but for the object: a human, lovable counterpart. Conceding, the analyst poses as an example by offering his own relationship functions, his empathy, his ability to retain and preserve the analysand within himself (holding) and productively change (containing) his conflictual material. The analyst's reflective, answering readiness to respond as a self-object of the analysand can also be used by him to build a continuous, coherent and positive Self.

These radically innovative ideas of the analytical process, the analytical stance and the contiguous technical concepts mark a new relationship between the analyst and the analysand, grounded in a two-person psychology, based on the theoretical turn to intersubjectivity.

The paradigm of intersubjectivity postulates that the analytical process produces a continuous matrix of mutual influence, whose phenomena can only be adequately understood as intersubjective interactive creations of the particular analytical pair. In this view, the analyst becomes the bearer of a subjectivity with which he does not show himself unreservedly but whose presence he may now acknowledge as legitimate and conducive to the process (Lesmeister, 2005, p. 30f.). The analytical encounter takes on the character of a non-symmetrical equivalence (Treurniet, 1996, p. 26), which is characterized by a higher degree of transparency, authenticity and reciprocity. Due to an intersubjective understanding of psychoanalysis, ways in which the Self interacts with its environment and how intrapsychic processes are coupled with interpersonal processes become apparent.

Psychotherapy: dialogue between understanding and knowledge

From the beginning, psychoanalysis is a dialogue, a therapeutic encounter at the language level with the goal of recognizing and changing the neurotic unconscious demeanour and complexes (see especially the following: Otscheret, 2005).

Dialogue as a philosophical method of knowledge consolidation is as old as philosophizing itself. It finds a first climax in Plato's Socratic dialogues.

We owe the discovery of the relationship function in dialogue, the discovery of the meaning of the 'other', of the 'you', the overcoming of the subject-object division, to the philosopher and Hegel student Ludwig Feuerbach (1804–1872). He realized that self-knowledge is only possible through the other person and in the encounter with the other (Jung, Chr., 2005, p. 228ff.). Continuing this line of thought is Martin Buber (1887–1965), who juxtaposes the objectivizing relationship of the 'I-It' with the dialogic encounter in the 'I-you'. His successor Emmanuel Lévinas (1906–1995) emphasizes the fundamental importance of others for our self-relation and relating of the self to the world. Jürgen Habermas (2005, p. 54 ff.) adds his ideas of a dialogue free of domination and communicative equality. Central to this thinking is the figure of the other and his otherness, which can only show itself from itself. (Lesmeister, 2005, p. 38f.).

According to Buber, existence takes place in the interplay between the functional sphere of subjectivity (I-it) and the respect or loving of the *sphere of the intermediate* (I-you). In the 'I-You' relationship lies the immediacy and reciprocity of the world of relationship. The *it* sphere, on the other hand, refers to the field of practical experience and of objects to use.

With regard to transference and relationship, the analyst's capacity for a sincere 'I-You' relationship in the Buberian sense is decisive (Jacoby, 1993). This presupposes that the analyst is able to regulate closeness and distance in such a way that he can empathize with the experience of the analysand and at the same time be able to psychologically reflect on the relationship. In this view, the transference relationship appears as an 'I-It' relationship, which, when the projections are withdrawn, can transform into an 'I-You' relationship.

The open dialogical principle in psychoanalysis deliberately allows for opinions, counter-assertions and contradictions. The relationship between the analysand and the analyst remains a special form of encounter characterized by asymmetrical thus limited reciprocity. If we consider the therapeutic consequences of intersubjectivity, this limiting condition for the therapeutic relationship must be taken into account as a sting of asymmetry (Otscheret, 2005, p. 79). An additional complication arises from the contrast between treatment technique and relationship, which is in constant danger of becoming blurred or perverted in the application of a relationship technique. This ambivalence of technique and relationship in psychoanalysis cannot be eliminated. Therefore, Lesmeister (2005, p. 55) demands a positive attitude of openness to this dilemma of difference, which paradoxically can create a symmetry in which the analysand feels understood in a healing sense.

Experiencing the relationship and the emotional availability of the analyst opens up a healing perspective. The basis for this is the willingness and ability of the analyst to access to the emotional life of the analysand in order to create meaning together (Orange, 2004, p. 25). Through the intersubjective genesis and anchoring of our Self, we consciously gaze at ourselves in our interactions from the perspective of a human counterpart. We acquire consciousness through self-reflection taken from the mirrored experience with the other (Altmeyer, 2000, p. 206), which is to be considered as different and equal.

Intersubjective recognition can be regarded as the core of a successful development of identity (Benjamin, 1996). Individuation as a lively process of organizing experience takes place intrasubjectively and intersubjectively. Out of the intersubjective experiences between relational persons arises an understanding of the multiplicity of the Self, which can allow intrapsychic asymmetry and contradiction. While it can endure ambivalence, it refrains from creating a unified, as it were, seamless consciousness. In every interaction, the Self is in multiple relation to the Other (Staemmler, 2015, p. 180ff.). The social interactions and dialogues of the past and the present constitute a current version of the dialogical Self. This may be real or imaginative and finds itself in dialogue with one or more others (heterodialogue) or with oneself (autodialogue). Our internal speech also has a dialogical structure in the communication of various parts of the Self or as an imaginary conversation with humans, animals, things. Inner speech serves self-regulation (Staemmler, 2015, p. 223ff.). Therefore, an important function of the therapeutic dialogue is the questioning and animating of condensed and stereotyped auto-dialogues.

The psychotherapeutic encounter becomes a dialogic process of mutual understanding at the point of departure where searching for conflict and meaning with the analysand preserves and respects his subjectivity, and a plurality of perspectives can be allowed. Meaning arises from the diversity of the parties involved, assuming that *tolerance for difference* can develop in communication (Bahner, 2002, p. 218f.). The dialogical principle extends the understanding and the handling of the transference and countertransference and allows a higher degree of symmetry and reciprocity in the analytic-therapeutic relationship. The central competence of the analyst is to shape the dialogue with the analysand in such a way that his inner working models of relationships and emotional expectancy patterns (Bovensiepen, 2004) can develop into mature relationship functions. Importantly, a process of unification needs to take place between the two – irrespective of whether a correct interpretation or an understanding with split-off soul parts is taking place (Blomeyer, 1995, p. 78).

In this process of unification, particular consideration must be given to the analysand's ego-structural characteristics, which have enormous influence on self-definition and on the type of conflict that can be experienced on the one hand and managed on the other hand.

Amplification: a Jungian way of psychotherapeutic meaning making

The dialogic method of amplification was originally developed by C. G. Jung and was initially used in psychotherapeutic treatment for the interpretation of dream symbols (Jung, 1946, GW 8, § 404). Through amplifications, an interpretive connection of personal conflicts to general human and archetypal conflict constellations is to be achieved. The Latin word *amplificare* means to add, enrich, reinforce.

Point of departure in amplification is an intentional and inner readiness to spiritually circumambulate an inner image or other psychic content. In the course of this process meaningful associations, including cultural and archetypal elements, are internally collected and made aware of (Cambray & Carter, 2004a, p. 125).

Thoughts, feelings and images capturing the analyst and felt in his or her countertransference (subtracting his own contingents), while in an internal demeanour of *rêverie*, are a symbolic or proto-symbolic expression of the unspoken and often not yet felt world of experience of the analysand. These take shape in the intersubjectivity of the analytical pair as an expression of the *Analytical Third* (Ogden, 1994).

Consequently, amplification is a sophisticated method of using analogies that can be used for psychodynamic understanding of clinical fragments, such as individual words, fantasies, dream images and even body sensations (Samuels, 1989, p. 38).

In contrast to the use of subjective associations of the analysand about his dream symbols or imaginations, amplifying refers to enriching the symbolic material with stories and quotations from fairy tales, myths, religious texts, with motifs and sequences from films, with examples from history, literature, from current political events and all fields of science. The intention is to link the analysand's symbols with universal themes of human cultural production, opening possibilities to anchor identity therein anew or further (Hill, 2010, p. 109). In this way, analysands can re-experience themselves as part of the human community and are able to put their subjective conflicts into perspective, i.e. to relativize them.

The Freudian School and the analytical psychology of C. G. Jung take different methodological paths in raising awareness of the unconscious and its effects. Freudian psychoanalysis is based on the fact that energy is released through the interpretation of events and conflicts that have been displaced into the unconscious or into the preconscious, which was previously needed to suppress embarrassing life events and fantasies leading to different neurotic malpositions.

Equally, Jung attached great importance to the fact that the analysand must become aware of his own participation in the emergence of his sufferings and complaints. Therefore, the analysand is mainly confronted with his 'shadow' at the beginning of treatment. 'Shadow' thereby refers to all that he rejects, criticizes about himself and refuses to believe, and therefore projects onto others, criticizing and fighting it upon the other. This gloomy 'shadow side' of oneself ought to be recognized – albeit often painfully – and integrated into the self-perception by way of withdrawal of the projections. Thus, the analysand can be acquainted via amplification with the riches of his unconscious psyche and with the symbolic content of human culture as a source of meaning-making available energy and creativity (Plaut, 2004, p. 157).

The classical treatment technique of Freudian psychoanalysis is based on the psychodynamic interpretation of the analysand's subjective associations or ideas about his symptoms, dreams, and interpersonal conflicts. The analyst meanwhile subdues to a technique of abstinence and steady held attention.

Subjective associations refer to the personal life story and the individual psyche of the analysand. These also include the experiences in the transference and countertransference situation. Subjective associations have a certain explanatory scope in that they can lead the way to a prehistory, to an initial conflict or to *complex* content that triggered them. In dealing with these subjective ideas of the analysand, the analytical process often stalls. After some time, essential childhood conflicts have been lifted from oblivion, and burdened insults have been remembered and are acknowledged. Important repetitive compulsions and maladaptive patterns have essentially become clear. Shadow themes such as one's own aggressive latency have been exposed and included in the psychotherapeutic debate. Grievances and burdens have not only been lamented but also mourned. Quarrelling that the lost is uncatchable and the past irreversible has taken place. Yet, despite all this the analysand can somehow feel 'misunderstood' expressing this openly through dream contents or through the manufacturing of resistance. An insufficiently illuminated perspective on the past could explain feeling misunderstood. Frequently, transgenerational conflicts, traumas or cultural complexes embodied in early caregivers and their imagines lay behind the problematic attitudes of parents who fail to sufficiently respect the individuality of their children. Parents' attitudes can include authoritarian ideologies, rigid religious systems, anti-women attitudes, discrimination based on migration or skin colour (Hill, 2010, p. 111).

Additionally, dissatisfaction could also stem from a lack of understanding sufficiently the future prospects of personal development. This might not have been sufficiently raised in the analytical process. In other words, the *quiet voice of the Self* wants to persistently and penetratingly be heard and not ignored. We assume that the emergent Self generates psychological necessities that want to be recognized by the conscious ego complex and strive for development in the individuation process and in resonance with relevant others.

The unconscious of the analysand offers access to the emergent Self through dreams and fantasies. Another access can be found via syntonic, unanimous countertransference of the analyst, who reacts to the inner world of the analysand with his own images, feelings, thoughts. The authentic need for amplification arises not only from subjective ideas of the analyst or the analysand. The impulse to amplify is above all a form of expression of the 'analytical Third', the interactive psychic field of the analytical pair (Cambray, 2001, p. 285ff.).

Subjective associations of the analysand necessarily lack linkages, which can be drawn through amplified *objectively* named associations. These connect the personal problem of the analysand with collective human themes and conflict constellations. *Objective* associations can be contributed by both the analysand and the analyst in a dialogical process. It is unique and touching for both analysands and analysts when such connections become visible and suddenly new perspectives and assessments of often complex problems arise. Through the discovery of meanings concerning the here and now and of humanity in general, the analysand accesses the unique individual-mythical experience of his own individuation process and the individual uniqueness of his type, his age, his personal history and his destiny (Neumann, 1978, p. 79ff.).

Comprised in the idea of self-development in a subjective process of individuation is the notion that inner changes take place again and again with the goal of making a previously unknown psychological necessity conscious by linking it to it. The symbolism of change is reflected, for example, in initiation rites or in religious ceremonial. Each transformation includes experiences of transcendence and mystery, symbolic death and rebirth as an expression of an ongoing dialogue between the ego complex and the Self. Change as a goal of psychotherapy is the psychological opposite of repression (Samuels et al., 1991, p. 236f.).

The potential for change inherent in the symbols of the analysand, as they appear in the dream material and in the complex topics, should be understood, used or integrated as comprehensively as possible.

The method of amplification opens up the possibility of making the conversion energy of emotionally effective symbols accessible and usable and of bringing alive the transcendent function. In the course of the process, the analyst will change as much as the patient will himself (Jung, 1929, GW 16, § 164f.), which is why the mutual unconscious connection between the analysand and the analyst must be examined continuously. Both images and symbols of the analysand and those of the analyst can be amplified. The cultural material acts as a *reference of relief* with which transference content can be enriched and understood more clearly (Fordham, 1974, p. 147f.).

C. G. Jung was not so interested in the fact that we have complexes, but above all in what the unconscious does with the complexes. The implicit connections between complexes shape intrapsychic and interpersonal relationship networks and the interactive analytical field through personal and collective-objective associations (Cambray & Carter, 2004, p. 123ff.).

Ideas and actions brought in by amplification can also be understood by way of recognizing the inevitability of the progressive emergence of transference-countertransference enactments as effects between the analyst and the analysand. Analytical enactments are dyadic active actions and incidents, which both experience as concerning the other. The ability to engage in enactments gives access to the flow of unconsciously constellated content (Cambray, 2001, p. 278).

The analyst influences the properties of this interactive field by actively offering amplifications or by merely *thinking* in an act of amplificatory introspection. Often even the discovery of a feeling that comes from an amplificatory introspection can be effective for change. It can also serve to maintain a prospective inner attitude for the development of the analysand.

Importantly, amplificatory ideas should serve the analyst to reflect on what is happening in the treatment room: whether, for example, a certain transference is in configuration or of which nature own illusory, perhaps projective, countertransference impulses currently are.

Analysts do not necessarily have to reveal their amplificatory knowledge. It is possible to use one's own amplifications indirectly in the sense of one's own intuition to get a feeling for where developments are going.

Amplifications can be described as spontaneous or deliberate, concordant or complementary, as temporary or accompanying psychotherapy over a long period of time, as internally unspoken or verbalized. Amplification can be done by the analyst in a calming, comforting or challenging way.

Spontaneous amplifications result from introspective countertransference incidents of the analyst. When I experienced stagnation in the analytical process for some time, I repeatedly experienced sudden physical sensations in the treatment example, such as feelings of warmth or cold, intestinal noises, irritation of the throat, sweating or an accelerated heartbeat. I remembered the image of endless desert or eternal ice, a place of starvation, and suddenly remembered photos of the polar explorers Amundsen and Scott. When silence has governed the session for some time I said suddenly: 'I'm thinking about the race of Amundsen and Scott to the South Pole.' As if electrified, the analysand replied: 'All this time I didn't feel you anymore – didn't one of them die there?' I said: 'Within sight of each other they both would have made it!'. This *moment of encounter* and the potential for change of the symbolic content of the small story marked one of the important turning points; 'visibility' remained a recurring metaphor in the dynamics of our relationship dynamic.

A syntonic-concordant amplification was the mythical figure of Parsifal in the later treatment example (see below), and his development from an ignorant, mother-bound fool without empathy to a compassionate Knight of the Grail, husband and father of Lohengrin.

A second, complementary amplification resulted from the Parsifal novel: Parsifal is also the stupid, unempathetic fool, thus a typical *Parsifal situation* representing one of not recognizing and not understanding.

Another, third amplification resulted from the 'psychotherapy' and transformation of Parsifal at hermit Trevrizent's place.

A further important complementary amplification became Goethe's ballad 'Sorcerer's apprentice' (Goethe, 1998, p. 276–279). One time the analysand had described himself as a *sorcerer's apprentice*, and he remembered its first verse from his schooldays:

Finally, away he went –
The old warlock!
Thus, now let his ghosts
live by my will.
I notice his words and works
and the custom,
and with strength of mind
I shall do miracles too.....

The exemplary analytical psychotherapy described in detail below, the following important and effective amplificatory carriers of meaning were initiating the discussion on parental complexes: the symbolism of the mother's bond reflected in the mythological developmental novel *Parsifal*, the great fantasy of the sorcerer's apprentice ballad and the oedipal individuation conflict in the analysand's self-ideal as a 'revenger of the disinherited' (Robin Hood).

Jung had introduced the method of amplification into his psychology as a hermeneutic or sensory method to make hard-to-reach or difficult-to-understand contents in the patient's dreams and fantasies accessible through comparative observation. In a letter Jung writes that his methodology compares to comparative anatomy or comparative religious history in how to decipher difficult ancient texts. As material for analogy he used texts of mysticism, alchemists and religions (Jung, 1989, Letters II, Letter to C. S. Hall of 6.10.1954).

Today, we would say that it is of great importance to recognize the *mytho-poetic connections* (Jung, 1961, GW 18/I, § 547) that form the interactive field of the analytical couple. In my therapeutic example, an important hero fantasy of my analysand was the guitarist Jimi Hendrix. His difficult and rebellious life story, his musical genius, his commitment and his closeness to death played an important role in his confrontation with the desire to be a hero himself, with his fantasies of grandiosity and his despair and suicidality. His former desire to become a working-class leader, an avenger of the disinherited, is reflected in Robin Hood's myth, his quest for meaning in the quest of Parsifal.

The goal of amplification above all is the interpretation of meaning: the symbolic side, the symbolic content of the language of dreams and the patient's ideas can be made accessible and insightful through comparative observation (Giegerich, 2012, p. 210ff.). Its method can be applied to any psychic material that can be understood and accepted by consciousness as an emotionally touching symbol. Besides observing the dream symbols, the aim is to develop a symbolic understanding of fantasies, hallucinations, imaginations, meditative processes, memories, interpersonal encounters and the relatedness to other real experiences. In particular, the symbolic contents of transference and countertransference incidents also require amplification.

It is our task as analysts to take up the symbolic side of the language of the unconscious and to convey it to the analysands, i.e. to develop and promote their *transcendent function*. It comes into action when consciousness can pick up a symbol, and by processing it, change takes place in consciousness. An important psychotherapeutic task is to amplify the perception of the analysand, especially when he runs the risk of ignoring significant symbolism (see *Parsifal situation*; Dieckmann, 1979, p. 176ff.).

Through amplifications we can make tangible those emergent, readily accessible collective and archetypal patterns of meaning in order to develop a common *language* and stories. Consequently, interpersonal and intrapsychic creative connections can be provided, of which the analysand can draw sense and meaning.

Amplifications can be handled in a similar posture to that of David Winnicott's 'squiggle game' (Winnicott, 1992, p. 66, reference by C. Caesar). The joint 'associative lead' of the topics discussed creates an intermediary space, a space of possibility in which an expanded interpersonal and cultural life of the analysand becomes available.

Let us once again recall that much of what analysands express is a projected personality. This means that we have to constantly reflect on who we actually embody. Relatedly, the analysand can be in transference mode even if we interpret the transference. Unconscious contents, and especially aspects of the imagines of parents and parent complexes, are inevitably projected onto concrete persons and situations. There is usually no way to explain or get rid of them, for example, through educational interventions. Conversely, this also applies to the analyst's transferences. Once aspects of the parents' imagines are projected, they inevitably remain so long until a process of change is initiated by both the Self of the analysand and the Self of the analyst.

We should be cautious and economical in the use of amplifications with younger analysands. Their yet insecure identity can promote the danger of being occupied and inflated by great mythological or religious symbols. For young men, for example, this can happen through the image of the hero or the Redeemer, for young women, for example, through the imagination of the erotic, male-dominating witch, as embodied today by artists such as Madonna or Lady Gaga.

Another danger is to use our amplifications defensively in situations when analysands come emotionally close to us and when we use amplifications to control our anxiety. Furthermore, it is to be warned not to support a depersonalization of defence processes with amplifications or to disguise a transference relationship in an amplifying way. It is continuously important to understand the countertransference aspects of amplificatory enactments, whether they are openly expressed to the analysand or silently noted inside. These and similar questions have to be asked repeatedly in the follow-up work at every session.

A blatant abuse of amplifications would be to impose mythological material to people and human relationships 'because it just fits the bill', or to want to see mythological figures 'incarnated anew' in people. In this way, the mythological material would deprave to mere jargon, voiding associations and amplifications of any metaphorical and symbolic value. Abusive would also be the narcissistic seduction of decorating one's own biography with the idea of some mythological figure being reincarnated in one's Self. Giegerich (2012, p. 212ff.) calls for a critical approach to such forms of simulating mythical experiences.

On the contrary, the depiction of amplification as set out above expands one's own space of thought in the analytical situation in a triangulating sense. In this way they can stimulate reflection during joint contemplation of collective motives, which then lead to personal connection and moments of encounter. Amplifications allude to a mythical archetypal quality and higher dignity of mental matter of fact (Bovensiepen, 2011, p. 302ff.).

Humour is a particular form of the helpful *Analytic Third*. Humorous contents can serve as complementary amplification countering real conditions (Freud, 1927, GW 14, p. 385) by creating a contrasting mood that often facilitates painful insights into the process of self-knowledge, while humourless attitudes usually intensify misfortune and suffering.

In analytical psychology, humour is found in the inner figure of *Mercurius* or *Trickster* with his tendency to cunning, amusing or malicious pranks, his ability to transform, his being at the mercy of all kinds of torture and his function as a saviour, who can transform the meaningless into the meaningful (Samuels et al., 1991, p. 222ff.).

My analysand's lamented renunciation of his plan to found a start-up company became easier in this sense when I could connect this problem at the right moment with Ringelnatz's ant poem:

In Hamburg two ants were trekking, their path lead to Australia backpacking.
    In Altona on the trail, their legs began to ail, so they did wisely renunciate the last part of their escapade.

(Ringelnatz, 1912. Pos. 16)

The *timing* of amplifications introduced into the flow of the analysand's personal associations is of great importance. Amplifications can be extremely artificial, disturbing or confusing if they remain mental foreign bodies to the analysand.

One aspect of the art of treatment that should not be underestimated is the analyst's empathy for moments in which the analysand is so emotionally accessible that intersubjective encounters arise through amplification. However, the observation might relieve that amplifications can have late effects by being initially simply heard and later remembered in the analytical space, then taking to effect.

In summary, amplification can be understood as assisting containment, anticipation and differentiation of the latent mental necessities for individuation processes in the therapeutic intersubjective relationship. Amplificatory stories let us rediscover the overarching human condition in our current personal life situation. Amplifications must be debatable for the analysand. They represent knots tying together elements of meaning in the interactive field of analytical encounter. In addition, they trigger motivational impulses, which can bring into consciousness previously formless internal images and experiences and organize them there. Finally, amplifications that are not governed by the analyst's narcissistic needs deepen the therapeutic relationship.

Linking amplification and interpretation

The inner work of the analyst consists essentially of the intellectual and empathic recording of the psychodynamics of the analysand's unconscious conflicts and how they are embedded in his ego-structural conditions, in his approach to the understanding of the symbols of the analysand's unconscious and in the preparation of the analysand symbolizing pre-symbolic experience for the analysand (containing). The results of the inner work of the analyst are co-formed by the unconscious communication with the analysand. In this respect, they also express an *Analytic Third* (see Fig. 2.1). As tools, the analyst's expertise, the use of his countertransference, his empathic relationship skills and his intuition are at his disposal to be able to imagine possible paths of development and individuation of the analysand.

The external work of the analyst is to communicate the results of his inner work to the analysand through amplifications and interpretations.

As described, the amplification serves to enrich the symbolic material, which can often be made understandable and meaningful and brought into the wider context of the general-human possibilities. The subjective associations of the analysand, for example, with certain dream symbols, then appear above the objective associations embedded in another, meaning-giving space of meaning (Figure 2.2).

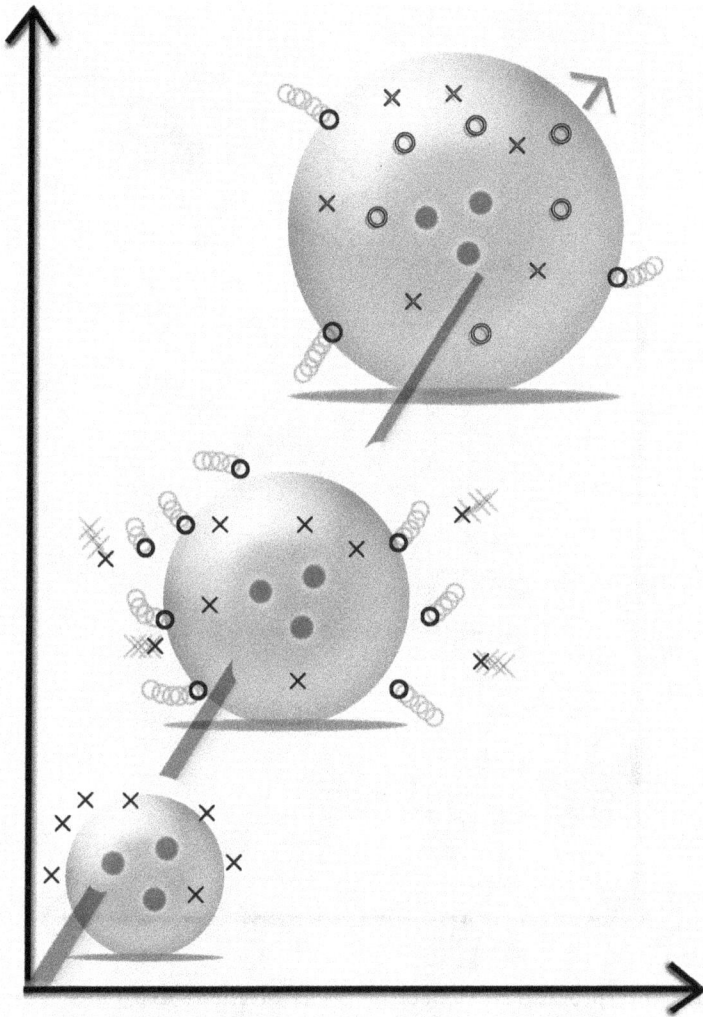

Figure 2.2 Amplifications: subjective (*x*) and objective (*O*) associations expand
the meaning space (spheres) of dream symbols (●●●).

The way of interpreting the symbolism of the analysand is different: through the process of interpretation, the symbols are to be traced
back to underlying complexes and conflicts. It is thus a process of reduc-
tion and focusing, the results of which are decisively influenced by the
psychoanalytical-theoretical orientation of the analyst (Figure 2.3).

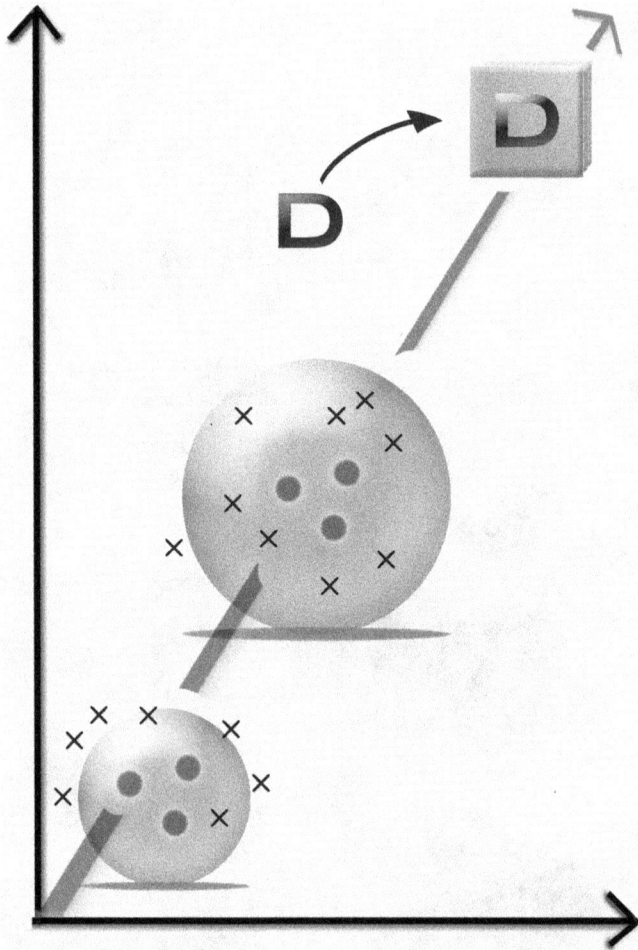

Figure 2.3 Interpretations: subjective (*x*) associations expand the sphere of meaning (spheres) of dream symbols (●●●), the interpretation (*D*) creates an interpretive space (cube), to which the dream symbols are attributed.

Amplifying and interpreting are thus complementary verbal actions in the flow and relational field of the analytic process. The analysand learns through amplifications that he 'knows more than he knows'. In dealing with the interpretations of the analyst, the psychodynamic depth dimensions are revealed. Both actions only become effective when they bring about emotionally valuable encounters.

Chapter 3

The Jungian psychoanalyst

Ethics of psychotherapeutic action and personality requirements

Our professional ethics requires every analyst to become aware of his inner coordinates and his psychotherapeutic 'belief system' through self-analysis and self-experience. Not merely enough, the analyst must also be prepared to apply this psychotherapeutic credo to himself in the sense of self-education. Jung summarized this in his ethical demand: '*You must be he whom you want to appear as*' (Jung, 1929, GW 16, § 167, italics taken from the original).

At moments when we discover problems and dark sides inherent to ourselves, which prove hardly open to the possibility of change, this task becomes particularly difficult for us psychotherapists. Additionally, we struggle when we feel too close a similarity to the difficulties of our analysands. In as much as our own personality is an essential factor of healing and transformation, elements of our character can also have an unfavourable and negative effect on the analysand.

Training analysis and supervision serve to develop a professional analytical stance towards the analysand, which should also be developed and internalized as an ethical stance and functional relationship (McFarland Salomon, 2011). We are facing an enormous suction in the psychotherapeutic setting, whether imaginative or real, which causes us to get merged into relationships and get entangled in projections, projective identifications and barriers in the ability to think. This makes it all the more important for the analyst to commit himself to put aside his own narcissistic needs while being concerned for the well-being and development of the analysand's individuality. The development of a positive relationship ethic of the therapeutic relationship is still in its beginning and is largely limited to the demand for abstinence and omission of various forms of abuse of analysands by the analyst.

The archetypal image of the wounded healer and the self-healing powers of the analysand

The idea that a healing person must himself be a sufferer can be traced back to the mythical teacher of the art of healing and of *Asclepius*, the injured, incurable centaur *Chiron* (Kerényi, 1997, p. 123). By analogy, we can say that the analyst himself must become aware of his own wounds or his own 'being a patient' if he wants to connect to the suffering analysand and his inner healer. Only if the analyst refinds own inner wounds empathetically-introspectively in the analysand he'll know what the analysand's missing. And only if the ill, injured analysand can project his inner healer onto the analyst will a treatment be successful and healing. Numerous evidence for this can be traced in, for example, placebo/nocebo research (Kächele, 2015). If the analyst is perceived as omnipotent and healthy in the treatment room, then the patient's analysand remains a sick person, addicted to the analysts' healing capacities, who is cut off from his own possibilities for self-healing.

The activation of the internal healer of the patient is made possible by the largely unconscious perception of the wounds of the analyst. The patient's inner healer unfolds his power of action when he can identify with the analyst and through this observe himself as a patient and reflect upon his own injuries. The psychoanalytic concept of the therapeutic ego-split refers to this connection (Samuels, 1989, p. 331ff.).

Importantly, within the therapeutic relationship, this identification becomes apparent when the analysand begins to express concern for the analyst or when he utters a concern of overloading him.

Considering a good analyst-analysand fit

Questioning a good analyst-analysand fit is an essential question both parties should ask during the initial sessions of a joint psychotherapeutic journey. This meaningful arrangement has far-reaching implications, which can more or less foster or even hinder vital developments for the analysand's future life.

In general, this fit is determined by the answers to the question, whether analysand and future analyst can engage in an affective alignment/harmonization or whether fundamental personality-related incompatibilities exist, which exceed the pair's analysable transference and counter-transference-related contents. In particular, even in an existent ideal emotional alignment, careful consideration needs to be drawn to an illusionary, idealizing transference, which could harmfully hinder the ability to think on behalf of both the analyst and the analysand. A spontaneous falling-in-love or an excessive fascination of the analyst during the initial meeting (Jacoby, 1998, S. 201ff.) could pose as example for such obstructive transference.

On behalf of the analysand, he should carefully consider if he feels he can develop a certain confidence that his complaints will be heard and taken seriously. He should receive a thorough diagnostic process with a view to his mental disorder and aspects of this biographic development, elicitation of all physiological conditions, his medical history, and all further past and present treatments and prescriptions. Furthermore, he should feel informed and involved in the process of seeking diagnosis and an appropriate indication and competently understand the formalities of health insurance cover and its regulations that help secure a safe and trustworthy framework for the upcoming psychotherapeutic treatment process. Within the initial meetings the analysand should also be under the impression to be welcomed into the sessions embracing his entire being, including all peculiarities, questions and doubts.

Frequently, analysands in search for a future analyst give extensive thought about the best age and gender of the future analyst. With regard to age I am referring to the analyst's ability to relate to the inner experiential world of the analysand despite possible large age disparities. In turn, as far as the gender of the analyst is concerned, it is worth asking whether the analyst might experience difficulties or conflicts with men or women in general. And if so whether working through them with the choice of analyst of that particular problematic gender would make sense. In case the future analysand is a member belonging to a particular minority group – sexually or ethnically – he should seek an analyst without sexual or ethnic prejudices. But it is also important to warn against analysts who react too enthusiastically or fascinatedly to sexual or ethnic peculiarities (Samuels, 2015).

An analyst should inquire whether he can find something appealing and agreeable and if within his counter-transference he can embrace the analysand with a primarily positive feeling and a hopeful treatment outlook. He should also give himself an account of narcissistic motivation in his decision to take him or her up as analysand. Such narcissistic motivation could relate to high-ranking analysands, physical attractiveness, idealizing utterances about the analyst or fine-grained gestures and signals expressing the willingness to be nourished and taken care of in any particular way and many more. The analyst should also ask to what extent his own shadow themata is addressed by the analysand's issues (Vogel, 2008, S. 108ff.). If unexamined or denied, the analysand would be harmed in his right to a broad and thorough psychoanalysis. Yet, if the analyst succeeds in removing the suppression of corresponding shadow themata, he must readily be prepared to dealing with his own character difficulties in what might become a strenuous personal process in the corresponding treatment. These themata usually refer to neurotic carriages that have 'survived' the analyst's own training based in trust analysis. Notably, those include narcissistic aspirations that seek to force the analysand to heal for the analyst's inner well-being (*furor sanandi*). Another common problem is the analyst's inhibition of aggression, which consequently leads to the avoidance of necessary confrontations with the analysand.

The willingness of the analyst to realize his own wounds and weaknesses in the treatment becomes an important prerequisite for later success. The potential benefit for the analyst would be that he too could personally develop in the course of attending to the analysand's injuries.

Objectives in the psychotherapeutic-analytical process

Developing a therapeutic relationship as relationship based on trust is deeply grounded in the shared formulation of aspired common goals. The analyst strives for the offset of a confrontation with unconscious material on behalf of the analysand. Conversely, the initial formulation of the analysand's goals is frequently referring to manifest psychological or psychosomatic symptoms and the desire for release and relief. Correspondingly, treatment often reaches termination when the analysand feels sufficiently advised with regard to his constellation of conflicts. Perhaps, he was seeking a relieving confession, a need to speak about a tormenting inner conflict, or perhaps he could discover previously unconscious content and connection and draw new motivation for life from it and develop new, somewhat more mature, attitude towards life. Perhaps his living conditions have developed positively through friendship, marriage, separation, change of profession, examinations and the like. Or he could discover a new practical philosophy of life for himself (Jung, 1944, GW 12, § 3f.). Or else, it could also be the case that the quota of hours granted by the health insurance expired and the analysand could not finance further treatment sessions.

Hence, the possibility exists that psychotherapy comes to an end while the analysand's engagement with the transformational aims of his Unconscious or his Self is still in full swing. Another important goal for the analyst should therefore be to prepare the analysand for the fact that through his psychotherapy, an access to goal-seeking mental processes is opened to him. And while these processes were unconsciously active and involved in the causing of suffering, they will accompany him also in his further life.

Chapter 4

Psychopathological concepts

Psychopathology and problems adapting to reality

An energetic take on libido

The foundation of psychic energy, or *libido*, in human beings is the tension reciprocally created by naturally given versus spiritual-cultural motivations. Freud considered libido as drive energy stemming from bodily sources which became channelled into spiritual-cultural achievements through sublimation processes. In Jung's view psychic energy holds universal character and is connected to the objective space-time-continuum (concept of *unus mundus*). Physics and psychology are complementary; therefore, psychic processes such as feeling, thinking, and sensing are all energetically grounded in a physiological basis (Jung, 1946, GW 8, §§ 440–442).

With this in mind, newer emergence theories and their application to complex adaptive systems attempt to shed light onto the complementary relation and the creation of meaning from *nothingness* (McFarland Salomon, 2013).

The theory of emergent qualities of psychological development assumes a continuous changing dynamic in sensing systems. They can drown in crisis of meaning or *chaos*, or they can forgo a structural change towards a more complex pattern with new qualitative characteristics, which cannot simply be explained from their original experience.

An example on the psychic dimension is *the occurrence of consciousness* as emergent quality of the central nervous system's development. A young infant's psyche proactively engages in with its primary caretakers and therefore directly influences the development of its brain structures. Further examples of emergent processes are later developmental leaps in self-confidence in the course of individuation processes. Accordingly, the sudden occurrences of psychopathological states can be attributed to emergent processes of human psyche.

Dynamically *libidinal-energetic and symbol-producing* processes underscore emergent processes of conflict resolution and problem-solving, which continuously create new patterns and meanings which psychological development or self-transformative processes can potentially avail themselves of.

The experience of libido is marked by a specific *strength or intensity* and an emotional quality (*affectiveness*) that transposes onto the targeted complex. This *affectiveness* is the basal layer of our personality. In this sense, our thinking and behaviour as well as our psychological values are the *offspring of our affectiveness*.

Intrapsychically, the energetic energy of our libido can be positively experienced as drive, affect, desire, wish, wanting, idea, performative motivation, impulse, urge, motivation, compulsion, conviction, attention. Negative effects can be experiences such as inhibitions, melancholy, depression, and lack of motivation. Here, libido is concentrated in networks of complexes, between which a constant flow of energy ensues. Within the psychic system this energy is in constant emotional and cognitive flow and change, which we can experience.

Inter-psychically, we experience the energetic effect of libido in all forms of attraction or rejection, which others evoke in us.

The energetic move is roughly directed towards goals. The energetic movement is directed towards either the future or the past, she can be directed towards to outside (extraversion) or the inside (introversion).

Future-orientated attitudes gain their direction through consciousness and our focus on the conscious demands of life. However, neurotic attitudes and effects of complexes may influence this attitude as well as unconscious individuation impulses, which occasionally lead us to seemingly *absurd* choices of action.

A regressive inner movement occurs when the conscious adjustment to reality fails. The occurring state of inability to make decisions or state of ambivalence can be understood as a congestion of the psychic energy flow. The longer the blocking of personal possibilities for conflict resolution lasts, the more *charged* the opposing viewpoints become. Consequently, a threat of dissociation arises, indicating a split in personality, from which the development of a symptomatic burnout syndrome or mania can arise.

With regard to the therapeutic relationship, the consideration of libido as psychic energy opens ways to consider thinking about where and with what amounts the analysand lets his libido flow: towards whom, in which interests and intentions, in the development of which abilities, in extraversion or introversion.

Furthermore, the energetic take on libido allows for the imagination of causes in the obstruction of flow of libido or its distraction from intended goals, for example through defensive actions by internal conflicts potentially resulting in the aforementioned symptom formations.

Finally, such an approach opens up possibilities of formulating the psychodynamic focus for psychotherapy based on depth psychology or a short-term therapy. The longitudinal view of psychotherapeutic processes allows for observation of ways in which libido flows transform and develop new goals, interests and passions.

Theories of neurosis

The intolerability of certain events does not suffice for the creation of neurosis, with its indispensable, compelling character of a phobic, obsessive or hysterical symptomatology. Historical changes in the display of neurotic disorder allude to their embedding in the super-personal Zeitgeist.

Psychoanalytic explanations for the development of a specific expression or Gestalt of a neurosis refer to the biographical points of fixation in the developmental stages of libido (oral, anal, phallic-narcissistic). Alternatively, they can, for example, be related to the idea of libidinal-sexual arousal being directly transformed into a fear neurosis. Causal trigger for the expression of neurosis is seen to be the re-experiencing of a supressed conflict, which the patient is unable to master since he is chained to a compulsion to repeat.

In contrast, Jungian psychotherapy repeatedly emphasized the potential inner meaning inherent in neurosis, insinuating an inherent value and a constructive function obscured by the suffering yet caring potential for development of the ill. The psychotherapeutic process ought to dedicate itself to the searching of the hidden function.

My experience teaches me from this view that neurotic suffering is predominantly tormenting and senseless and should by no means be mystified as a path to healing from crises of meaning. Particularly interesting in this context are Wolfgang Giegerich's (2014) following considerations, according to which neurosis is a free, productive creation of a soul sickened in contemporary history, since modernity has stripped the human psyche of its formerly encompassing and absolutely compelling metaphysical and religious dimensions.

Neurosis as the work of a sick soul besieges a person, takes possession of him, virtually constructs itself out of the pieces of personal biography and the life's circumstances of an individual. Neurosis is the expression of the modern disease of the soul, its content entirely senseless and the agony of its suffering without any value for development or individuation. Within a demystified modern world, the soul simulates an inclusiveness, an absolute, via neurotic symptoms, for which the suffering human has to pay the prize.

He primarily experiences a self-alienation, self-division: something within him confronts him, dominates him, is withdrawn from his disposal and eludes all efforts to regain one's own sovereignty. In the first instance, mental illness in general and neurosis in particular represent a disturbance in one's relationship to the Self: partial psychic functions escape one's control and become independent in the shape of neurotic symptoms, such as compulsions, depressions, panic attacks, disorders of impulse control, ego disorders, hallucinations and so on. They confront the human who hosts them, deprive them of their autonomy and control them.

Changes in the ways the Self is experienced influence the course of the disease, for example, through negative self-perceptions, self-reproaches and self-assessments. This becomes an essential component in the progression of the disease in terms of negative relationship to one's Self – frequently to a point of suicidal intention.

Equally, mental illnesses are relationship disorders because the responsiveness, the ability to respond in social contexts, is impaired. As a result, disturbances of communication in private relationships and at the workplace occur, which, conversely, impact negatively on how we experience ourselves (Fuchs, 2011).

The task of psychotherapy therefore consists firstly in helping the suffering person to regain his independence and emancipation from feeling overwhelmed by the staging of the soul. Secondly, therapeutic efforts are concerned with the treatment of all those conflicts, traumas, intolerabilities that cluster around the neurotic or psychotic complex nuclei. Therefore, it envisages the elimination of symbolization disorders and the improvement of ego-structural and identity disorders. Thirdly, negative maladaptive relationship patterns and their vicious circles must be clarified and changed. Scope and depth of these processes of clarification and change depend on the status and quality of the therapeutic relationship and its viability.

Disease models in analytical psychology

Complex theory is not only a theory about the general functioning of the psyche, but it is also an original theory of psychopathology of analytical psychology. As described in 'The complex psyche' section of Chapter 2, complexes can influence the ego complex in such a way that the personality changes in the sense of a complex identity. This becomes evident when thinking about the effects of power complexes, inferiority complexes or fear complexes. Complex theory as a model of disease is concerned with intrapsychic aspects pertaining to identity conflicts by way of the emergence of psychopathology from the effects of co-existing complexes competing for influence on the ego complex.

The dimension of relationships in psychopathology, i.e. the question of how intrapsychic conflicts and structures affect interpersonal and thus also therapeutic relationships, is examined in complex theory from the perspective of experiences already made, which have found their expression in the implicit memory of relationships and in the network-like organization of unconscious complexes.

Actual interaction dynamics can be measured with additional instruments such as Operationalized Psychodynamic Diagnostics, OPD (Arbeitskreis OPD, 2009).

Causal factors of diseases in analytic psychology

In summary analytical psychology assumes the following intrapsychic psychopathological factors to be effective:

- congestion of libido in the network of complexes,
- influences on the ego complex by other complexes,
- acceptance of a complex identity,
- regression of libido to previously inhibited complexes,
- pathological functions of the ego complex and typological problems.

Mental illness and dysfunctional behaviour become manifest when the ego or the ego complex can no longer fulfil its control functions in the grip of neurosis. Such control tasks face internal conflicts, on the one hand, and structural characteristics such as self- and object perception, the regulation of one's own affects and the regulation of relationships on the other.

Ego consciousness can be weakened in its decision-making ability if it cannot dissociate sufficiently from unconscious self-identity: one's personality then falls prey to competing autonomous complexes. The unstable experience of interpersonal relationships of a borderline personality may pose as example in this case.

If someone is solely identified with the ego, the Self remains limited in its effect and in its impulse for individuation. Consequently, existential crisis particularly in midlife can result, leading to depression and somatic disorders.

If the ego complex is not capable of imaginative dealings with other complexes, the differentiation and integration of these complexes cannot take place. For instance, a thought and action guiding desire can attempt to enforce the experienced lack of parental attention by ways of an imaginary attention by the inner images of the parents. Since one has not sufficed parental expectations in the way one has naturally been, many professional and relationship choices might have been taken against one's own abilities and chances leading to perpetual calamity.

Conscious intentions are diverted into senseless actions, if the ego is not able to control the *problem function*. For example, when feeling represents the inferior function, sudden and self-damaging decisions are imminent in numerous situations in life.

In case conscious ego fails to integrate the *shadow*, addictions or deviant and exploitative behaviour can arise. Prevailing of a one-sided functional style of ego consciousness together with a connection between fantasy and pathological symbols can lead to the development of narcissistically or psychopathically disturbed personality. Furthermore, when ego is temporarily or permanently overwhelmed by other psychic contents, such as affects and imaginations, and towards which the ego complex is too weak to maintain the unity of the individual, a psychotic breakdown or a psychotic dissociation of thinking and feeling can occur.

Various psychopathological symptom formations can also arise within psychotherapeutic processes and represent special challenges for the therapeutic relationship. It is particularly important for analysts to evaluate the limits of what can be achieved together with the analysand. Psychotic disorders of a schizophrenic or of affective nature and acute suicidality regularly require psychiatric drug co-treatment or an interruption of psychotherapy for the duration of an inpatient stay. With regard to drug treatment, medical psychotherapists should delegate the simultaneous drug treatment of their analysands to colleagues, unless there is a clearly defined reason from the peculiarities of a therapeutic relationship.

Neurotic conflict, structural qualities and psychosis

Defining for a *neurotic level of a mental disorder* is the persistence of an intact unity of personality and ego identity despite relative autonomy of pathogenic complexes. Essential for a neurotic development is a weakening of ego-complex control. This means a relative, meaning still connected, rupture between ego and a force relying on unconscious content, which has been coined counter-will. Neurotic processes are accompanied by entirely orderly psychic content; no blurring of conscious ideation takes place.

Structural disorder, in contrast, highly limits the affective functions, as well as those required for differentiation, whose necessity include the regulation of self and its relationships (Rudolph, 2005, S. 48ff.). Consequently, a more or less pronounced insecurity of identity can pursue.

Early disorders can be the result of inadequate emotional and cognitive care and stimulation by the caregivers. They can also be an expression of traumatic experiences, such as violence, child abuse, forms of deprivation, severe illnesses and congenital physical impairments.

In any case, early disorders affect the development of a person's *structural qualities*. In the case of early disorder, the world is experienced as seeded in archaic and frightening interaction experiences due to limited possibilities for reflection and comprehension, partly relating to the circumstance that persons with early disorder can only make limited use of their symbolizing function.

Structural qualities are mainly anchored intrapsychically within implicit relationship memory. Easily triggered intense negative affect obstructs the thought of a consistent benevolent, positive interaction with other persons. In this regard, early disorder can be conceptualized predominantly as continuous and pathogenic disorders of interpersonal affective coordination abilities. Subject to the pressures of easing the integrative function of the ego complexes, the whole psyche can split *dissociatively* in various complexes.

In *psychosis* complexes of the whole psyche create dissociated fragments, which are often irrevocably detached from functional contexts. The occurrence of psychosis is hereby less a reflexion of specific unconscious content but rather an expression of the weakness of integration of the ego complex or an impairment of its inhibitory function. The vulnerability-stress-coping-model illustrates these interrelations.

The effects of complexes have more profound consequences, if a schizoid vulnerability already exists, than in the case of neurosis: affective experiences, such as strong emotions, overpowering fears, emotional imprisonment in delusions, frightening disintegration of logical thinking, and other factors characterize the specificity of schizophrenia.

The figure of dream life experience somewhat resembles that of schizophrenia, which regularly has been characterized similarly to a great dream in its intensity of experience. In contrast to dreams, schizophrenic psychosis is accompanied by an unsystematic fragmentation of conscious ideas. This is the difference between a schizophrenic complex and a normal affective complex. While attention and concentration can be strongly hindered by a normal affective complex, it never destroys its own psychological elements or contents as a schizophrenic complex can do.

There is little doubt of the subjective meaningfulness of psychotic content and delusional systems helping the afflicted to orientate himself in the world surrounding him. Delusional content need to be accepted and taken seriously in *psychosis psychotherapy*. This specific method of psychotherapy helps to comprehensively understand the patient in his perception and aids to guide his subjective thinking back to realistic states of his social reality and social relationships (Jung, 1914, GW 3, § 408ff.).

Psychopathology and individuation

Ego and Self

At times of crisis, after his separation from Freud, Carl Gustav Jung had been confronted with a steady and relentlessly threatening inner stream of images, which he captured in both word and image in his *Red Book* (Jung, 2009). Various figures of his imagination and dreams progressively began to form an order representing typical human possibilities of experiencing and behaving, which he later called *archetypical* or *timelessly given*. Thus, he compared them to instincts.

Furthermore, he found that the mentioned stream of images inheres a compelling tendency towards progressive integration and a forward movement towards a goal, which seemed to be stemming from a *configurative centre* in the unconscious, which Jung described as *Self*, contrasting it to *ego*. He named the entire process, fundamentally, biographically centring self-development or *individuation*. In Jung's experience of a higher Will the experience of *Self* coincides with an experience of God, as can be found in archetypical inner images and figures such as Mandalas, Christ and

Buddha, corresponding to the contents of the collective unconscious (Jung, 1911a, GW 6, § 456f.).

In Freudian psychoanalysis, Kohut introduced the term 'Self' to describe a holistic view on the psyche. He defined the Self as goal and centre of the psychological universe in contrast to self as content of a psychic apparatus or a self-representative of the ego.

Jungian theory, namely Fordham (1985, S. 90f.), developed the important understanding that Self emanates from an original, *primary Self*, hence marking the beginning of the child's development. Under the idea of a primary Self we can see the foetus and the newborn child as beings separate from the mother. Such primary Self is believed to hold a differentiating albeit rudimentary *archetypical sense of otherness* and therefore of Self from the very beginning. Winnicott's term of the *true Self* (Winnicott, 2002, S. 193) as creative experience of being alive as well as Bion's concept of *O* (Bion, 2007, S. 46) approximate Fordham's view of Self as developmental point of departure.

An early perceptive ability and a primary joy of conversation concretize the *sense of otherness*. From the beginning we are dialogical beings and the experiences drawn from our interactions and conversation determine the continuous development of the content of the implicit relationship memory. Our inborn activity to seek out objects is connected to an early competence to draw caregivers into social interaction (Dornes, 2000, S. 19ff.). The infant's ego begins to organize itself *through togetherness*. It becomes interested, ready to act and emotionally oriented towards the non-ego, referring to the world of objects, the world of others.

The development of the structure of Self reaches a peak in establishment when the ego can understand itself as object and thereby reflexively relate to itself. Relatedly, the opening of an inner mental space and the beginning of a conceptual lexical-symbolic imagination or representation of one's world of experience can be observed as of the 18th month of life in the form of an increasing ability to psychological regulation, mentalization and the availability of self-reflection.

The structure of Self and the structure of relationships to others mature in intensive interweavement. In contact to the caretakers guiding and organizing functions of the ego differentiate from inner images of Self as well as from images of relevant others. Development of self-autonomy is facilitated by the safeness of attachment relationships.

An important step in individuation of the becoming Self is achieved when an *autonomous Self* is experienced, which has developed a *feeling of identity*. Furthermore, the person is able to regulate his self-image and self-worth as well as his ability to control and act in relationship. He is emotionally in condition to communicate to the inside and the outside. It is such a Self that experiences *self-efficacy* (Arbeitskreis OPD, 2009, S. 116).

When I feel complete and in harmony with my life's goals, I can assume to be in touch with my Self. The self-regulating process of individuation is longitudinal, during the full course of life dependent on the ego complex and its integrating function.

The concept of intersubjectivity, which incorporates the otherness, strangeness and difference of the other, exceeds the idea of Self being a solely egocentric and omnipotent psychic unity. The intersubjective perspective of Self and individuation acknowledges the other to uphold his otherness as a counterpart without me having to become identical with him (Benjamin, 2002, p. 128ff.). In this view, Self can experience and welcome diversity, differences and contradictions in his relationships to various others.

Current Jungian discussions evolve around the model of Self, grounded on a dynamic systems theory, which denies a pre-existence of a template or structure for both the mind and behaviour. This model interprets the Self as a phenomenon that develops by itself. Therefore, it voids the assumptions of a primary Self. According to this viewpoint a Self is created via dynamic patter of a complex system, which includes genetic, physiological predispositions of the infant, ascriptions of caregivers with regard to the infant's intentions as well as cultural and symbolic influences designing the surrounding of the developing Self (Hogenson, 2004).

The emergence model outlined above postulates an entirely epigenetically driven development of Self and denies the existence of a genetically predisposition of primary Self. Nonetheless, observations of numerous actions performed by the infant appearing highly competent and an unmistakable individuality as from the point of birth point towards the idea of every child already being born as a person.

In conclusion, it ought to be said that the intentions of the Self are not always benevolent. Our decision shall never fully rely on the voice of our unconscious, which is nature and thus beyond good and evil. The vigilance of our conscious is always vital (Jacoby, 1985, S. 105).

Within the therapeutic relations, the analyst is called to particular attention to the states of the analysand, whereby he becomes flooded by impulses of the Self, that do support individuation, but are yet too immorally nature-like. Coming-close-to-oneself in the analytic process can be at times accompanied by intense feelings of happiness or expansion, leading to previous relationships and love partners of the analysand to appear nearly pale or colourless, thereby mobilizing fierce impulses of separation. It would be a devastating therapeutic mistake if the analyst would become infected by the enthusiasm, without considering and defending the containing long-term meaning of the analysand's love relationship and his social environment.

Regression and progression

In Jungian psychoanalysis the term 'regression' refers to a process of mental regeneration, a temporary pause, an inner regaining of strength as opposed to an understanding of a return to earlier stages of development. If difficulties arise in individual life which cannot be solved with the existing coping strategies or if the creative connection with the unconscious is disturbed, the libido must turn inwards in order for further development and progression to possibly pursue.

The following case study exemplifies a seemingly insoluble contrast between the analysand's striving for autonomy and the duty to assume responsibility in fatherhood. Subsequently, severe regressive depressive and anxiety-influenced conditions showed up, including acute suicidality. The appearance of the dream symbol of a Djinn, a genie in the bottle, representing an effect of transcendent function (see 'The integration of creative expressions' section of Chapter 10, Fig. 10.4) presented a carrier of meaning, which led the analysand out of his former psychological impasse and made new life decisions possible. Arising from this symbol of the unconscious, he suddenly possessed will power and an idea of possible goals.

A regressive introversion into a chaotic depressive state can lead to sudden meaningful events in spiritual life via symbols carrying the function of transformation, since on the level of the collective unconscious, implicit order exists, which we can access via the function of symbolization (McFarland Salomon, 2013, p. 237ff.).

Symbols of transformation

Wondering how chaotic, polarized states of suffering and ailing conflicts can lead to new order and new meaning led C. G. Jung to the discovery of the *transcendent, symbolizing function*, whose image-like form of expression aids the discovery of new and useful pathways of solutions. These imply the dissolving of inner contradictions to new levels of understanding.

Symbols are carriers of meaning, which collate an image and a context of meaning. Symbols are so to say *pregnant with meaning*. They are potentially the best and most comprehensive expression of possible webs of meaning or hypotheses. Hence, symbols are first and foremost subjective. If *social symbols*, such as religious or political ones, capture entire groups of people, their charge of meaning can have an enormous effect. Once this charge of meaning has been apprehended and exhausted, symbols lose their impact power and continue to hold only historical relevance (Jung, GW 6, §§ 896–908).

The ability to symbolization appears uniquely human in its moving power, and we avail ourselves of an enormous treasure of symbolic images of the world and ourselves (Dorst, 2015, S. 21ff.). As laid out in The 'Alchemy, transformational processes and psychotherapy' section of Chapter 2, the human mind holds an archetypical preparedness to create symbolic and emotionally moving images.

Underscoring the psychic transformation as an *emergent process* is the idea of our human ability to continuously create new mental patterns and meanings, thus allowing access to self-reforming and self-healing powers (see the 'An energetic take on libido' section in this chapter; McFarland Salomon, 2013, p. 222ff.).

From a Jungian perspective the conviction of the reality and power of the symbolic is crucial. In light of the emerging processes involved in the formation of ourselves and our lifelong individuation, we can assume that the symbolic implies more than just a system of representations but a relatively autonomous system of web-like self-organization.

An important hint in this direction comes from the patterns of time in the 'creation' of dream symbols during REM sleep. They appear to be invariant of scale and thus a universal trait of the psyche. They seem to follow similar mathematical structures as language acquisition, the course of the Nile or the firing of neurons. The hypothesis that the world of the symbolic manifests nature-like and that our psyche is provided with the ability to perceive symbolic phenomena offers us a further pathway of understanding the relationship between dream and reality. Dream symbolism of the analysand or amplificatory impulses of the analyst could be a crossroad of the realization, of change of phase from a continuous process of symbolization to comprehensible and evocative patterns of meaning (Hogenson, 2005).

This holds the following implication for the analytic relationship: that the analyst can implicitly or explicitly communicate to the analysand his soul's ownership of an eternal source of symbolization processes, which only a severe brain injury could dry up (Kaplan-Solms & Solms, 2000, p. 48ff.). The confrontation with the strange and mysterious intelligence of the unconscious psyche can lead to an enormous expansion of ego consciousness and creativity.

Chapter 5

Treatment in analytical psychology

Healing through self-awareness: paradigm of individuation

Individuation is an intersubjective process of self-discovery aiming at the differentiations of personality and one's awareness of an inner Self. We virtually become human only in relation to other humans. In this way, a developing Self is already in its foundation at least a twosome, *dyadic* unity, containing *transcendence* in form of dialogue with an 'other'. In the course of individuation, the abilities develop for dialogue, demarcation and relatedness in the course of a biographical *becoming*.

Ideally, we would perceive an individuated individual as one well in touch with himself, *as* him as he is and less as imitating others. He would treat others as they truly deserve. He appears original and authentic. His behaviour would cast an image neither too self-confident nor too insecure. He would be free from pressures to be too individualistic or too extraordinary. His self-confidence would be stable and not based on the suppression of fears and doubts. He would possess a solid sense of his viability and could feel relatively comfortable and secure in various social situations and circumstances.

To fulfil this ideal in one's life is difficult since its formation is underscored by cultural and historical influences and reflects our *cultural complexes*. Self as historically burgeoning is always *generational*, hence seizing a particular historical space in which it organizes collective and individual experiences (Bollas, 2000, p. 231ff.). Consequently, we ourselves are history. Individuation is subject to a dialogue with a burgeoning society and ensues in argument with a social reality. Merely, if we relate *socially responsible* with such reality can we develop ideas of a shared and human future (Odermatt, 1998).

Once we can accept the idea of our becoming and having become in our relationship and in our history, we can also accept as unexceptional, however unique, our own process of individuation and that of our family, friends and analysands. Every human is at any point in his life more or less consciously entangled in individuation to the point of individual predispositions and possibilities. Everyone has a varying genetic makeup, varying psychological processing possibilities of past experiences and varying experiences with a supportive, obstructive or traumatic social environment.

A child is born with innate and, in this sense, archetypal perceptive and interactive abilities: with the ability to localize signals, with spatial perception and a sense of colour. It prefers structure to disorganization. Additionally, a child is stimulus-hungry, likes novelties and is primarily designed to interact with people or to evoke interactions. From the first day on, the newborn learns from its experiences and forms expectation patterns. It gives preference to the human voice, human face and smile over all other stimuli. Towards the end of the second year of life, all essential building-blocks of individuation are available: body control, self-awareness and distinction of what is internal and external, language, active interest in other people, an evolving conscience with ideas of good and evil as well as the use of symbols (Samuels, 1989, p. 205f.).

Individuation is a lifelong process, with changing psychological baselines causing different foci during different phases of life, such as childhood, latency, puberty, adolescence, early career, marriage, family building, mid-life, elder age (Dieckmann, 1991b, p. 171). At any stage authentic self-experience is possible, however much they might be aided or hindered by caretakers, teachers and partners. In this regard every time and every developmental stage offers adequate space for individuating. This in turn requires the *Self* to be able to change dynamically and, one may postulate, develop stemming from an original (primary) *Self*, which has always been existing next to a motherly *Self*. Originally consisting of archetypically based possibilities, the *primary Self* is able to confront and align the archetypically available expectations with outer and inner reality through an unremittingly process of utterance. Pursuantly, the encounter with reality in the eyes of the other becomes internalized as self-experience.

On the condition that the confrontation of archetypal expectations with the outside world does not exceed the respective ability of the *ego* to integrate, ego consciousness arises and expands. The developing ego is in a concordant and phase-appropriate process of loosening and finding its inner cohesion.

At every stage of life, the ego is confronted with both archetypal possibilities and images that have not yet been realized, as well as objects of the outer world that are still unknown. Inner images, called *imagines*, are composed of elements of archetypes, defensive operations and real objects of the outer world. Ego development and Self-development are inextricably linked. Hence, individuation is the specific developmental and maturation process of that particular baby, toddler, adolescent, adult or of myself embedded in a relationship context.

Individuation is like a river in which I swim: a flow of experiences, encounters, adventures, thoughts, dreams, fantasies and meanings of people and things. This all may be passing or else it remains in me as an idea of otherness, subsisting entangled in my development. Those incidents that affect me are self-expression and self-education at the same time and openly accessible to my reflection.

Life provides different individuation possibilities and tasks at different times. Ways in which I cope with them impact on earlier and future effects of individuation. The intersubjective perspective emphasizes the context-dependency of the intrapsychic, from which the individuality of the Self emerges. It is the 'Ego-You-relationship' in the Buberian sense that creates an intermediate, an intersubjective emotional and relational field through which ego differentiation, differentiation from the collective norms and differentiation from the contents of the collective unconscious can unfold. Individuation is always connected with a social task, either through the passionate pursuit of a task for society or by lovingly turning to the other: '...that as a lover I love the person through whom I receive God's gift...', as Jung put it somewhat pathetically (Jung, 1916a, GW 18/II, §§ 1104).

Individuation processes can manifest differently in different stages of life – or they don't manifest at all, because they are completely inconspicuous and part of everyday life. We might only be aware of one resulting effect: a harmonious state, a balance between a person and his environment. A vital criterion of individuation is the development of humbleness and the assumption of one's own limitations.

Interpersonally, despite a one-sided or incomplete, and to a certain extent limited individuation, individuals can be experienced as *constant partners*, perceived without ambivalence and placed a large importance in someone's life. The very touching closeness to disabled pupils in inclusion classes may serve exemplary here.

Developmental processes reach consciousness mostly when the *flow* of individuation is hindered or interrupted and life's demands cannot be mastered, thus requiring a re-orientation, such as is the case after arising of symptoms or during crisis of identity.

Individuation, both as process and result, is never free from possible restriction or damage, be it because shadow elements are not reaching resolution, conflicts cannot be resolved or structural feature is insufficient for coping. At such point of crisis analysands seek our help. An intersubjective perspective on individuation relating to psychotherapeutic intervention targets mapping out, fostering and accepting both commonalities as well as differences. As a part of this, the creative curiosities of neurotic, psychotic or borderline perspectives need acknowledgement by the analyst.

In our forthcoming case example, the analysand initially appears hindered in his personal development in such a way as if his individuation process has come to a halt. From the beginning the question of how to reignite this process and which direction the analysand needed to be accompanied has significantly escorted and guided the treatment.

Since there is no 'normal case' for individuation processes, one's own counter-transference in relation to one's own guiding principles had to be considered and observed, i.e. also the confrontation with one's own values and with one's own ethics of life.

Two treatment methods are particularly fruitful. First, the repeated reflection of the regression level of the analysand summarized by the inner question: which person, of what age, with what injuries and with what abilities do I have to deal with at the moment? Is there a child, a teenager or an adult sitting there? Second, a repeated swap from the diagnostic to the empathic perspective: How did my counterpart feel in this or that situation? How, for example, was he seen by his father or girlfriend? What wishes and hopes does he place in the therapeutic relationship right now and in the long term?

Uroboros and dragon fight: integration of parent complexes

The inner acting of parent complexes and the contemplation of their influence on ego and Self are of vital importance to the development of personal autonomy and self-efficacy. A person's ability to resist the power of parent complexes when needed is an extremely important step for individuation.

Symbolically this power is represented in the image of *Uroboros*: a snake biting its own tail, forming an imaginary circle – a yet undifferentiated masculine-feminine figure – rendering an escape from encirclement necessary for girls and boys not to remain *subjugated father-/mother-children*.

Accordingly, the analysand from our case example said during one hour at the beginning of the analysis: 'I have always tried to present a certain picture of myself, but it has nothing to do with me; it is very deep. I always want to arrive as a beaming hero, but this is a facade....' Behind this façade or *persona* he usually experienced a profound fearful paralysis and depressiveness, which stood in complete contrast to the image of the shining hero. His heroic ideal expressed an expectation or a future-oriented goal that he could not yet satisfy: heroes in the beginning carry only the *images* of the future.

The dragon fight of individuation is a metaphorically necessary fight, whereby the hero fights against the old system of rule (Neumann, 1999, p. 144ff.). Therein, two father- and two mother-figures are of particular importance: the *evil king*, image of the personal father figure, hopes the son will perish in the battle against the monsters, the power of the unconscious and the uroboic Great Mother. Great fears must be withstood in this fight, whereby the *good mother* stands by the hero as the heroine and as a sisterly virgin (see section 'The integration of creative expressions' of Chapter 10, Figure 10.7). Mythologically, the hero in a patriarchal system was victorious with the help of the *good father*, the God-Father. Due to the independence as mark of modernity, he often has to take up the struggle alone, as Prometheus, and has to be able to think alone and ahead, in order to overcome the transpersonal, collectively shaped imagines of the father and the mother.

In his interpretation of the fairy tale 'Amor und Psyche' (Neumann, 2004), Erich Neumann describes the tasks and the path of female individuation as a feminine heroic path of *active* lovers. Angelica Löwe (2015, p. 329ff.) uses the text to show how affection can arise through an overturn of aggression following persecution and self-founding. Psyche's path follows the narrative of the 'unsolvable' tasks: with courage, perseverance and faithfulness, she pursues the goal of overcoming the separation from Eros.

Jungian psychoanalysis fosters an understanding of these forces at play as transpersonal powers, since it evidentiarily shows how frequently we tend to identify them wrongly and excessively in the real parents.

Consequently, the therapeutic relationship should recurrently open up perspectives of parents as real persons, with both strengths and weaknesses. Furthermore, it aims at developing an ability to empathically engage with the situation of parents at various points in one's life-time and to be able to recognize the parents against the backdrop of the grand- and great-grandparents' generation. In our case example of the analysand it becomes increasingly visible how much his parents had to endure the agony of an authoritarian grandparents' generation. In this way he could begin to *feel* a solidarity with his family, of which he only *knew* before: how his mother's father took care of her and her three children after the death of the analysand's father.

Shadow integration

The analysand presents the intolerability of an inner or outer experience in a symbolic staging via his psychopathological symptomatic manifestation – an inner and outer symbolically contextualized plot. Our ability to recognize the analysand and his symptomatology in regard to his relationships and living conditions opens up a potential for the unfolding and understanding of meaningful latent interconnection between that which is suppressed and that which is denied manifesting itself through symptoms and suffering.

A common observation is the way analysands become ensnared in maladaptive relationship patter and repetition compulsions, thereby repetitively restaging their yearning and desires as much as their failing. An essential and repeated problem poses the inability to-be-alone-with-oneself, either as various forms of obsessional pegging to another person or in loneliness and isolation potentially culminating in catatonic states or suicidal behaviours. Such pegging becomes more or less related to the analytical situation. Thus, Jung claimed a central goal of treatment to be the analysand's chance for experiencing an inner footing in the personal unconscious in the course of psychotherapy with the aim of being able to be alone without feeling lonely.

An important step in this direction is the confrontation with the *shadow* as content of the personal unconscious. *Shadow* hereby implies all elements a person fears or despises in himself. However, the *shadow* should not be seen simply as evil, since a dark side is germane to being essentially human. Importantly, the analysand should identify his problems relating primarily to his own shadow and to a lesser extent to others, for example parents. Confronting one's own illusions is evidently painful. Additionally, the concept of *shadow* comprises all that is not yet and may never be conscious in the darkness of our unconsciousness.

In considering one's own shadow, one deals with the underdeveloped areas of one's Self on one side or with damaged, blind, atrocious ones on the other, which have been suppressed, denied or split from one's social *persona*. The shadow appears in dreams, projections, transference and countertransference. It evokes fears via phantasies and nightmares in images such as a murderous persecutor, hideous and grotesque mugs, repulsive and perverse male and female figures. Confronting such images can be impairing and destabilizing and represents a particularly difficult and delicate task in analysis, which, specifically with structurally impaired patients, needs to be consensually limited. Frequently, such patients perceive development and change for longer times less as liberation but as threat to their identity and their psychological existence (Roser, 2011, p. 212ff.).

Working through shadow aspects is exacerbated by the fact that shadow content is usually *projected* onto other people and personal relationships. Furthermore, negotiating shadow aspects is difficult, since they initially cannot be discriminated from archetypical content of the collective unconscious. In the process of becoming aware of shadow elements, this archetypical content emerges or is *heaved up*, arousing enormous fears due to their strangeness. Such archetypical fears may be represented in mythological figures, such as the black, death-procuring Indian goddess Kali, with her crown of sculls and a dead child in her arm, or the zombies in the increasingly fashionable genre of horror movies and series. Such images remind us also that life is tragically limited and that the unconscious is not necessarily good and benevolent (Marlan, 2010).

A general problem is that shadow content can never be fully conscious, since the conscious recognition of certain content can cause subduction to the shadow of others. Figuratively, the spotlight of my consciousness may shine on a certain shadow topic, such as the beguiling narcissistic seducibility through praise and affirmation, while simultaneously other more primitive impulses, such as wanting to procure praise and affirmation by violence, become faded out and shuned in the shadow. Consequently, the shadow holds a significant moral and ethical problem: its plaguing content is in no way tolerable, in need of change and yet required to be accepted and integrated.

While dealing with the shadow, the ability of orientation to the inside – 'being alone with oneself' – is vital, since introversion facilitates the negotiation with the content of both personal and collective unconsciousness, aiming at differentiating oneself from collective norms, collective consciousness as well as from collective unconscious. In the course of the therapeutic process, violent reactions of transference can occur, since the claim for individuation itself calls for resistance and rejection of conformance to others, including the analyst (Jung, 1916a, GW 18/II, § 1094).

The analyst bestows the role of an elucidating, reassuring and affectionate *soul guide* in relation to these resistances and their means of expression. He should make destructive elements of his transference, splitting and projections visible to the analysand and simultaneously offer a safe therapeutic relationship. Pursuing, both the analyst and the analysand can establish connections between inner personal images, phantasies and conflicts with corresponding collectively human ones. Encountering archetypical images through *amplification* can then become bearable and thus facilitate development. Then, cultural creations such as myth, literature and symbols can be relayed to personal issues of the analysand's life (Hogenson, 2004, S. 74f.), ending in accepting the necessity and possibility of *shadow integration*.

Peter Schlehmil poses as an exemplary, a man who sold his own shadow to the devil. Free from shadow, Schlehmil paradoxically ends up being swathed by the biggest psychological shadow: by believing he can buy himself everyone and everything. In the end, despite his riches, he is made an outcast by human society and is prevented to love (Chamisso, 1923).

In sum, confronting shadow themes of the analysand is a delicate and difficult task for the analyst and can potentially threaten the therapeutic relationship. If the relationship is not sufficiently solidified, dealing with shadow- and shame-content can lead to the abruption of psychotherapy. This is even more likely when ego-structural problems, such as a narcissistic mode of defensive formation, are being answered *morally* and implicitly condemning by the analyst. In our case example some problematic aspects of the analysand's shadow and shadow integration are detailed in the 'Shadow recognition and shadow integration' section of Chapter 9.

Development of the transcendent function – moments of meeting

C. G. Jung used the term 'transcendent function' (1916, GW 8, §§ 131–193) to describe the ability to understand the interrelation of initially conscious and unconscious intrapsychic material in such a way as to achieve a broadening of consciousness. Developing this ability in the analysand ought to be mediated by the analyst. Yet, the quality of the development of this ability within the analytic process depends on the cognitive potential to insight of both parties and can be understood as an interconnected regulatory process in a shared analytical field. Thereby, the analyst's quality of the experience with the transcendent function and his personal symbolic stance are vital for the developmental potentials of the analysand, as long as his own possibilities of introspection and thinking about himself and his interaction with others are in the making. Important for the development of a shared process of *becoming aware* are *moments of meeting*, which manifest out of the implicit relationship knowledge that both the analyst and the analysand hold internally. These processes have been studied in detail by the Boston Change Process Study Group (Stern et al., 2002; Stern, 2005). They describe ways in which the interactive field of analysand-analyst develops towards a common extension of conscious understanding – touching and changing both via moments of encounter as transactional events that enrich the implicit object knowledge of the analysand. This process towards insight is often felt as a new experience of the *Analytic Third*, perceived as a touching and exhilarating experience of the unfolding Self and of a lively creativity (Carter, 2010). The *transcendent function* is the technical expression of a progression towards finding a piece of identity, aiming at integrating one's relation to oneself.

Working on ego structure

The integration of transgenerational influences and traumata

Mental distress, such as melancholy, enduring depressive mood and anxiety syndromes, void of an identifiable conflict or a biographic trigger often relates to transgenerational unconscious distress. In our case example both grandparents and the father of the analysand sympathized with national socialist ideology; both grandfathers were very authoritarian, both police officers, such as the father has been for some time as well. Imagines of both grandfathers and the father coin the father complex in this case with dark and highly threatening aspects affiliated. Jung wrote (1931a, GW 17, § 93): "...one ought to articulate the phrase, that not parents but their family trees ... are the true creators of children and explain more of their individuality that the immediate, random so to speak, parents."

Such observation led to the conclusion that the engagement with a transgenerational perspective with the analysand and the goal of integrating the imagines of the ancestors are essential integrative steps. Firstly, it becomes meaningful if an analysand can perceive himself as standing on the shoulders of his ancestors. Further then, this can lead to illuminate problems, faiths and shadow topics the ancestors fought in order for the analysand to exist today, having prepared his social starting point. Another access would be to become aware which talents or other mental or physical aspects the analysand owes up to his ancestors.

Treatment of biographical trauma

As laid out above, fear reactions require only one triggering exposure to become inerasably connected and inscribed in the *fear system* as a dangerous thing. In our case example the father was a tall, strong man and was very loud. The father has often bellowed furiously. '*A patriarch.*' The family shivered of fear in his presence: '*Father was the castle's keeper. Sometimes he came down to his bondsmen, mother and us children, to obtain tributes. Or: Father rides his motorcycle up the porch: a humungous appearance.*' He rarely spent time with the children, and there has never been physical contact or conversation. His father was strict with him and wouldn't tolerate any discussion. Until his brother was born his father was relating to him, and after that his brother became his father's favourite.

Treatment of mentalization deficits

The implicit relationship memory itself has direct repercussions on the further development of the brain but also on all future relationships and 'conversations'. For this reason, the impact of *early disorders* on the development of a person's *structural characteristics* is immense. Early disorders are an expression of limiting and damaging *mentalization conditions*. Mentalization means acquiring the ability to distinguish between inner and outer reality and the capacity to think about both. Early and complex mirroring processes by the mother as well as engaging in as-if-modes via play with parents, siblings and other children breed the acquisition of this ability to mentalization (at around four years of age) due to aligning the ideas and feelings of children with the reality and inner shape of others. Traumatized individuals frequently lack this experience of play and parallel their inner world and those of others, mistaking it for reality. Their inner feelings thus become experienced just as to be outer and real occurrences (Fonagy & Target, 2006, S. 369ff.). Interactional experiences for individuals with early disturbances therefore becomes determined by overwhelming fears. Easily triggered and powerful negative affect can make a 'positive interaction' barely imaginable. Social situations easily lead to what C. G. Jung has coined 'introversion of libido' (Jung, GW 5, §§ 448–450) and to which Melanie Klein's 'paranoid-schizoid position' (Klein, 1972a, p. 144ff.) refers to according to the object-relation perspective.

Subsequently, early disturbances primarily signify ongoing disturbances of the abilities to interpersonal affective attunement. These abilities are re-presented intra-psychically by specific shaping of the imagines of others and by the complexes of implicit relationship memory. Furthermore, early disturbances are accompanied by severe impairment to contemplate relationships and conflicts. Individuals whose inner working models don't represent such contemplation are incapable to extract value and symbolic meaning neither in their own actions nor in that of others.

One access possibility to understanding affected analysands is to comprehend the content of unconscious communication between the analysand and the analyst. The observation of one's own countertransference hereby becomes the organ of understanding: the perception of one's own images, impulses, emerging fragments of thoughts, body sensations and the pursuant inner amplification of these perceptions by the analyst. His own countertransference responses are the result of his ability to symbolize, demonstrating his reflective function. This function can stipulate a psychological meaning to the analysand's actions that the analysand himself is incapable of recognizing (Knox, 2004, p. 79).

Attachment disorder

Infants hold an inborn tendency to seek closeness to a trusted person. A healthy adult tends to attempt satisfying the infant's needs. The *attachment system* represents an independent system of motivation, aiming at creating physical *closeness* to a caregiver and emotional *safety*.

Resilience against biological, psychological or psychosocial developmental crisis develops from an interplay between genetic factors and the environment, whereas the regulatory *early attachment relationship* carries a particularly high compensatory significance for genetic vulnerability factors.

Various attachment types can be distinguished, whose quality relies on the *tactfulness* of the attachment person in early life. *Attachment quality* is measured within a spectrum of secure attachment to avoidant attachment. Securely attached children show distress when separated from caregivers and joy upon their return as well as seek their comfort and are easily calmed, following which they return to their play and exploration. Avoidant, even disorganized, attachment is marked by the children's inability to handle attachment-related stress. They display little pro-social behaviour, are aggressive or withdrawn and attempt to handle difficult situation on their own (Rass, 2011, S. 34ff.).

In order for adult relationships to succeed and parenting competence to prevail, attachment experiences are paramount, since they lead to attachment security, tactfulness, empathy and an ability for affective resonance. Attachment experiences play a crucial role in individuation processes by influencing ways in which we shape our personal selves. Additionally, they mould our intentionality in making decisions during crucial periods in our lives and when confronted by choices (Lesmeister, 2009, S. 258f.).

Attachment system and the wish for a satisfying, interdependent and realistic relationship remain effective throughout life. I consider of enormous importance the possibility of the analysand to experience a trustworthy, emotionally attentive, constant and crisis-resistant relationship with his analyst, which at the same time is well-defined.

Chapter 6

The therapeutic space

The therapeutic space in which we encounter the analysand has several dimensions, real and imaginary, on his behalf and determined and arranged by therapists. The analysand comes with his expectations and previous experience. In general, he expects help and support in overcoming difficulties that are no longer bearable to him and that he has failed to resolve. Analysands usually come up with problems whose origins they seek in themselves and for which they hold themselves responsible. They come with an inner imaginary picture of the psychotherapeutic situation. They expect to meet a competent helper, somebody who would understand their problems, a counsellor, a healer, a guru: in short, someone who is more competent in problem-solving than themselves. With the help of his psychotherapeutic methods, the therapist should be able to change their mindsets and attitudes in such a way that their ability to overcome conflicts and own coping possibilities could improve.

Hence, analysands come with an inner image, a *psychotherapist imago*. All the more often this represents well-known pictures of Sigmund Freud or the idea of meeting an elderly wise lady. Furthermore, analysands come with ideas about a adequate psychotherapeutic room and, for example, a certain piece of furniture – such as *the couch*.

The assumptions of therapists are usually complementary: We expect that a person seeking aid will come to us looking for help based on our competence and knowledge, willing to learn in the broadest sense. We expect the analysand to come to us with confidence, granting us the advance of trust, enabling him to speak openly and undisguisedly about his own problems, and to leave the protection of his everyday caution towards us, thus overcoming his limits of shame. Further, we expect the analysand to rely on the psychotherapeutic conversation to be conducted under protection of confidence and confidentiality. We trust the analysand to feel safe and secure through our benevolence, our ethical obligations towards him and our professional competence.

We offer a shared meeting place, which essentially communicates ourselves as well as about our perception of psychotherapeutic work by way of its furniture and arrangement. It is advisable to occasionally try to observe one's own treatment room through the eyes of an analysand. Looking at the entire room, we can ask ourselves: 'What have I set up for myself and what have I set up for the analysand?' 'What serves my self-portrayal and what do I offer the analysand, so that he can feel well and comfortable on "his" armchair or couch?'

Encounter and setting

The analytical meeting takes place in the analyst's practice room. To and from it, the analysand follows an individual pathway, which he will follow regularly for a long time, perhaps even years. The time spent on this route often gains meditative character in preparing the coming session or an inner follow-up on the one gone by.

Not only the person of the analyst, but also his room with all furnishings gain special qualities and characteristics in the analysand (Guderian, 2007). The conscious perception of the practice room or any contained detail, such as an image, for the first time can be regarded as anxiety-reduction in the analysand. The process contains an inner appropriation of space, where every small change can be seen and potentially criticized. Spatial appropriation entails making restitution: towards the end of the analytical process, the room is 'given back'. In the treatment example Mr. M. said at the end: 'I banish you and the analysis to the end of the world.' Space-relatedness as well as the relationship to 'own's' armchair or 'own's' couch deserves attention and can serve to explain the analytical relationship and the analytical process.

Sitting or lying down?

There is no 'standard procedure' in Jungian psychoanalysis. This means that neither the frequency of the sessions nor sitting or lying down during psychotherapy is given a more than functional meaning in individual cases. These aspects of the setting should be discussed with the analysand and agreed upon as changeable rules. It is advisable to give the analysand the freedom to arrange himself on 'his' couch as he pleases.

The Jungian analyst will usually position himself in such a way that the patient lying down on the couch can make visual contact with him. Invisibility is not required.

The question of whether to be treated in an armchair facing one another or lying down on the couch has gradually become a question of the analyst's personal style. I ask my analysands to arrange themselves on the couch so that they feel comfortable there and have their position towards my seating.

Frameworks and rules

Jungian psychoanalysis, like all other directions of psychoanalysis, assumes a secure framework and reliable appointments between the analysand and the analyst to be extremely important. The entire setting, the space, the time, breaks and holidays, even the body of the analyst and his appearance have containing, holding functions. Many symbiotic needs of the analysand are temporarily projected onto the setting. Such needs often only surface when the setting is changed or broken up (Wiener, 2015, p. 471). The framework includes all time agreements, such as frequency of sessions, fixed weekly appointments, duration of meetings, vacation regulations, cancellation regulations for sessions that the analyst cancels and sessions that are not attended by the analysand. The framework also includes comprehensive clarification of the (in this case German) process of health insurance's refund policies, state-aid and subsidy allowances, application review procedures and subsequent continuation applications in the approval process as well as their overall quotas of the amount of total sessions granted by public or private health insurances. In all these questions, an *informed consent* between the analysand and the analyst should be sought.

Finally, the framework includes the absolute certainty for the analysand that the analytical relationship is maintained by the analyst under all circumstances and protected against any misuse.

The healing asymmetry of the psychotherapeutic relation

The uniqueness of the relationship between the analysand and the analyst can be described as a special form of encounter characterized by *limited reciprocity*. The relationship remains asymmetrical because it has a specific goal: it wants to promote the encounter between the ego and the unconscious of the analysand and enable inner mental differentiation and promote awareness. In this unique respect, it is both natural and unnatural. It can create intimacy and is professional at the same time. Entangled in this particular relationship, it is important that the analyst can be used by the patient's Self.

This relationship experience with the analyst and his emotional availability allows for new experiential qualities of being implicated with another and relating to that person, which can strengthen one's own self-esteem. The basis for this is the analyst's willingness and ability to immerse himself in the emotional life of the patient: "Psychoanalytic understanding means creating meaning together" (Orange, 2004, p. 25). All of the analysand's existing experience has already been given meaning and has been interpreted by him cognitively and emotionally according to his possibilities.

Psychoanalytic understanding needs to relate to the analysand's memories as fertile ground for change: the emotional memory contains the quintessence of one's past relationships; one's life story rests in one's whole being. Growing insight and emotional understanding of the analyst can alleviate pathological effects of life events of the analysand and make them gradually controllable and tolerable.

Focus on current conflicts

The biographical background is seen as a history of experience and development in the present-day conflicts. In Jungian psychoanalysis and psychotherapy, the investigation and resolution or management of the analysand's present conflicts is of decisive importance. These should be captured as precisely and empathically as possible and be openly present in the treatment room. The earlier genetic conflicts are only of interest to the extent to which they influence the current conflict situation as either unconscious psychodynamic conflicts or as particular pathways of individuation of the analysand.

For Jungian psychoanalysis, the decisive factor of neurosis lies in the present, not in the unresolved conflicts of the past. Neurosis is created and staged every day; it is part of our present-day personality. Concordantly, it is also overcome and healed in the present day. To what extent the detour via the biographical past is necessary must be carefully clarified in the discussion with the analysand (Jung, 1930a, GW 4, §759, 1934a, GW 10, § 363). In my experience, analysands usually assume that earlier relationships and experiences are decisive for their current conflicts.

It is therefore only of limited importance what a life-historical conflict or a respective traumatic experience has caused and set in motion and where the analysand's earlier attempts to cope with it have led. Understanding the psychodynamics of the actual conflict, symbolically expressed in the present symptoms, refers both to the desire for development and to those inner obstacles that stand in the way of its realization. This attitude relieves analysands, particularly when their life-historical memories are fragmentary and difficult to access and is understood as presently experienced *persistent conflict*. Such conflicts need to be distinguished from *Current conflicts*. *Current conflicts* are experienced as temporary conflict-laden stress, which is conscious and accessible for immediate processing and resolution (Arbeitskreis OPD, 2009, p. 96).

Diagnostic perspectives

The understanding and description of the actual external and internal conflict constellations should lead to the question of what the ego and the Self of the analysand struggle for in these conflicts and what they defend themselves against.

In contrast to the 'history' of the analysand, which goes into the clinical diagnosis, it is a matter of starting from his experience, from his 'story'. Are there discernible major conflicts, developmental disorders, maturation delays and structural constrictions, and are development resources visible?

By seeing the analysand as himself in a certain chapter of his path of individuation, a perspective opens up into the past. His self-constitution (personality, identity, structure) and self-experience (self-awareness, satisfying and conflictual relationship experiences, life achievements) can be associated with his developmental conditions, his social environment, hindering and facilitating genetic and epigenetic factors and the relationship context in which he was born and grew up.

With future developmental opportunities in mind, the following questions can be asked: Is there an unconscious meaning in the experience and behaviour of the analysand, which we classify clinically as neurosis or personality disorder? Do experiences and behaviours mainly serve the patient's defences, are they primarily to be understood as defensive, or do they also contain development potential, even if they are hampered?

Chapter 7

First contact, indication, treatment

Case studies should describe the theoretical orientation of the practitioner, his epistemological and methodological approaches as well as his understanding of change and relate them to the problems of an analysand and the course of his psychotherapy. If possible, case studies should be instructive – and debatable. This presupposes that thorough insights into the course of treatment are given both longitudinally and in various cross-sectional descriptions of the situation. For these reasons I have decided to restrict myself to the presentation of one psychoanalytic treatment instead of case vignettes from various psychotherapies. All essential treatment principles apply equally to long-term psychotherapies based on depth psychology and short-term focal psychotherapies based on depth psychology. Different therapeutic variabilities within different procedures are explained in the 'Indication and dialogical processes' section.

Pursuantly, I would like to show how by approaching the reality of my analysand's life both empathy and a shared process of understanding emerged from the analytical encounter itself and mutually mirrored interpretations and explanations.

Introduction to the case study

The following case study offers an insight into the course of a Jungian psychoanalysis undertaken according to the principles and methods described above. Particularly important in this treatment were creative suggestions of the unconscious psyche of the analysand through significant dreams at important turning points of analytical process as well as creative impulses in the form of artistic ideas, such as sketches and drawings. The duration of the psychotherapy lasted two and a half years and included 280 individual sessions. Literal quotes from the analysand are shown in italics in this and the following chapters. All personal information is as carefully anonymized as possible; I name the analysand Mr. M.

I am particularly grateful for Mr. M.'s trust and agreement to use his dream material and his creative creations.

Initial contact, diagnostics, indication and treatment request

A tall, lanky man hesitantly and slightly bent forward walked through my door with a tentative smile, greeting me with a heartfelt handshake that surprised me. Mr. M. was dressed like a student: black leather jacket, striped jeans and sneakers. He appeared insecure for a 37-year-old man. His stubbly beard and nasolabial wrinkles made him appear older than he was.

He came with the car he drives as self-employed delivery driver. During our initial sessions, he kept a brimmed wallet on his lap. He later said: *'Money becomes a part of me; it represents the manifestation of pain; I can't let it go.'*

He remains enrolled in university but for a long time ceased to attend lectures. However, it is important for him to 'be in'. He expressed an unhappiness with the way he portrays himself: *'I appear too timid, lazy and dishonest.'* Further, I sensed a latent angry undertone in his voice: *'The doctor who referred me should have prepared me for the long commute to your practice!'*

His language was guarded, his emotional expression monotonous and retrained. However, his deep sadness and despair filtered through to me and could only be felt by himself at a later point in treatment. I felt sympathy and compassion and have been reminded of a friend of mine. After a short course of conversation, I have been determined to proceed with this treatment due to my spontaneous sympathy and the impulses of my countertransference.

He described a situation: *'I stand with my back against the wall, don't know where to and where from; all is a huge mess. I live on distant goals, don't know if I can love and cannot keep on living this way. I want to understand, gain clarity and finally escape this mist in my head.'*

He turned to his GP after a crisis with his girlfriend led him to increasingly feel depressed, doubting himself and tormented by intrusive thoughts of death. He and his girlfriend, a slightly younger, recently graduated lawyer, have gone through an abortion, which triggered the crisis.

During the time of pregnancy, he felt himself permanently contending ways of escaping the enormous *feelings of duty* and *thousand constraints* a child would bring to him. Since this, his relation to his girlfriend is ruptured: *'The lightness is gone, and I have become a fanatic of contraception.'*

He often retreats in his apartment for days, lingering in daydreams. The only other thing he clings to is his work: *'This work and my apartment are my niches.'* His daydreams are coloured by large professional success or by self pity: *'They will only wake up when I am dead!'*

Current symptoms

He complained about depressive moods, lack of motivation accompanied by an unbearable feeling of paralysis, pondering the meaning of life, thoughts and feelings of guilt, helplessness, hopelessness and aimlessness in relation-ships and profession. He understood these feelings as an expression of the inability to clarify his situation mentally.

Additionally, he experiences himself getting lost in small things, he senses fears whenever he has to deal with men in particular, his ability to sexual experience is impaired and he drinks too much alcohol. He feels somehow afraid to commit himself to anything, thus everything is always 'in the air'. Furthermore, he often has thoughts of death and about death entailing pleasant peace.

Initial trigger for his anxiety symptoms was the abortion of his master's thesis, which he could not complete due to an *absolute misjudgement* of what was realistic and his own expectations; it was supposed to become a *stand-ard reference book*. This point marked the end of his flight of fancy – the façade was broken. He understands his current conflict situation as a con-siderable intensification of the problems that have existed for over six years, since he now struggles to feel up to the task of taking responsibility for his partner and a possible child. His main reason to undergo psychotherapy is to become internally free in order to give meaning and direction to his life.

Burdens from biography

He tells me about strong nocturnal states of anxiety during childhood ac-companied by phantasy of someone or something threatening him from beneath his bed. Anxiety exacerbated after entering Kindergarten and pri-mary school. During adolescence they subsided; however, a worry of break-ing down remained.

His first depressive reaction he traced to was when he was 4 years old, after the birth of his brother. His earliest memory, possibly corresponding the former: *'I crawl alone in the street carrying a cigar box which belongs to my father collecting glass splinters – something beautiful. I am being badly hit by a bicycle and injure a whole in my eardrum. I continue to crawl but now I am deaf.'*

Another depressive reaction he remembers at the time of registering for Kindergarten. He felt dumped there after the birth of another brother. He felt again like this after entering a new school class once the family moved from northern Germany to Bavaria in the south.

In his initial primary school he has been known for causing unrest, clown-ing and disruptions. But he has been integrated to some degree into the class community. His new school in Bavaria received him well. However, by fault of his mother's intervention a boast was revealed, and filled with shame, he has withdrawn himself socially for years from his schoolmates. He became a lone wolf and an outsider. The only contact he upkept outside of his family was with an *outcast* in town, with whom he built and exploded small bombs and prowled through woods and meadows: *'I became a poacher.'*

His father's death caused by an accident inside the house during his 16th year of life was particularly drastic. The father had suffered irreversible carbon monoxide poisoning, and the son arrived as the father was dying.

Typological considerations

Due to the influence of transference and countertransference, the typological constellation remains relative. Transition between thinking and feeling, sensing and intuition are fluid as is the transition between introvert and extravert attitude (Blomeyer, 1979). Hence, the pursuing considerations on typology remain relative, meaning valid only in relation to me as analyst.

I experience a main function in the analysand, which he reverts to in order to easily and naturally orientate himself – this is his sensing. This becomes apparent in his artistic expression and sense of aesthetic sense, which he, for example, applies to furnish his apartment. At the beginning of treatment this dominant function was peculiarly lifeless and interpersonally uninhabited, while it was very well usable and alive in his perception and artistic creations.

His feeling function was energetically particularly beset and conflict-laden. It had to be understood as problematic function, leading Mr. M. repeatedly into withdrawal, isolation and proximity to death. The feeling function has also been the slow function, often in need to express itself through somatic or via physical paraesthesia.

Partially developed was his auxiliary function thinking. While he did often complain about a lack of clarity and transparency when thinking within his field of tangible consciousness, he progressively in the course of treatment felt delighted by his thinking brightening up and becoming more realistic.

He frequently achieved intuitive creations leading to countless projects, which he then dropped again. Regarding other people he intuitively saw rather their negative and problematic sides in them. The course of development of his professional path, later leading to his successful application and hiring, was driven intuitively. A statement I made parenthetically during one of our sessions became a *key statement* to Mr. M., helping him to turn towards his intuition with more awareness: 'You don't trust your intuition.'

His attitude has initially been extravert; he observed everything others brought towards him with great attention and lived predominantly reactive. In this way, his conscious energy flowed mainly towards to objects. This was particularly manifest and expressed in his inclination to look out for and make bargains. Introversion was initially effective beneath the conscious threshold and during treatment developed towards himself asking *what is personally really important to him in his life.*

Symptom-related complexes, ego complexes and archetypes

The onset of the anxiety symptoms can be understood as an internal regression movement. The negative character of his mother complex, wanting to cling firmly to him, prevailed with all its force for Mr. M. during onset. Paralleling the way in which his mother imago, or inner image of mother, wanted to shackle the son to her, he became afraid of being shackled in much the same way by his girlfriend. Reactively, an archetypal counter-image originating in puberty became stronger in the unconscious: the idea of remaining an eternal youth, a *Puer aeternus*. This became attractive to Mr. M., and he broke away from the idea of starting a family, refused any further responsibility for others and became unequal to his professional advancement.

The following details the negative effects of the mother complex described above: Mr. M.'s current anxiety-inducing situations in the field of consciousness are the conflicts about closeness and commitment with the girlfriend, culminating in her desire to have children and sexual claims. The strictness of the mother, behind whom the authoritarian grandfather on the mother's side can still be seen psychologically, justifies his fears of authority. Impersonalized expressions of these fears can be found in compulsive avoidance of *contamination*, e.g. through female hair in his bathroom. Further anxiety-inducing situations in his prehistory were the physically overwhelming and seductive presence of the mother and the beating by the father instigated by the mother. At the complex core the mother mobilized the background figure of *the Witch* with incredible power. She could, for example, bring his first love to immediate extinction through her degrading involvement.

The negative effects of the father complex appear complementary as follows: competition with men generally and currently triggers fear in Mr. M. Furthermore, there are the fears of authority as well as noisy men taking up much of a room and the fear of sexual failure.

The father complex comprises all experiences with the personal father and grandfathers, as well as archetypal aspects, such as male activity and phallic aggression, and images of man as sire and destroyer. Behind dream images of black-robed judges (see below) and 'inquisitors' appears the Nazi and police background of father and two grandfathers, who embodied the negative senex figures in the space of the father archetype (Jung, 1971, p. 96).

Fearful situations stemming from his biographical experience between the ages of 10 and 20 included the family's move to a foreign place, his exposure as boaster in his new class, bullying and exclusion by the classmates, the beating by the teachers granted by the parents, the fear of the discovery of onanism as well as the horror of pubertal homosexuality. In addition to this were memories such as the one that his father had forced him to wear girls' gym shoes in gym class, which he had bought cheaply as a special offer, or his father directing a hairdresser to a haircut that was against his will.

Any physical closeness was fended off by his father. Within the father imago his sound volume, body-size and fatherly force are present together with the image of his perfectly polished black shoes and the razor-sharp creases.

Particularly formative was the shocking accidental death of his father accompanied by his last words: 'Take care of your mother!' This event led the then 16-year-old to immediately and without great sadness live up to his father's last will: he did so by clothing himself in his father's clothes and shoes, which were still far too big, and later wanting to ride his motorcycle. The tragic death of his father inhibited any revolution, any uprising against the father.

It had to be assumed that the archaically oedipal experience of the real death of the father, or 'patricide', had traumatically fixed the image of the father as a law in his personal habitus impeding any differentiation of consciousness within Mr. M. In his dreams, black robed judges appeared over and over, judging and ruling over him. Consequently, the liberating myth of revolt (Jacoby, 1975, p. 536f.) remained obstructed towards the father and a shadow fantasy regarding one's own murderous violence could not be differentiated.

Finally, inner resources emerging from positive aspects of his parental imagery included the strength and accuracy of the father, who had recognized his drawing talent, and the inherent trust of his mother in him as well as her optimistic outlook on his life.

Indication and dialogical processes

Ahead of treatment an important decision had to be made in shared agreement: whether his treatment should primarily be oriented towards a pathway of pragmatic conflict- and problem-solving or whether his grave and life-threatening symptomatology mirrors a hindering of his personality development, thus requiring an analytic treatment approach working predominantly in transference. He asked for my advice after I laid out the various possibilities bestowed upon us in health insurance-covered guideline treatment options.

Conflict and relationship underscore two poles of psychotherapeutic work: the interactional character of the unfolding of primary impulses of transference in the now and the character of interpretation founding the development of intersubjective-libidinal constellations of transference constituting the long-term therapeutic process (Treurniet, 1995, S. 114).

Even though I have chosen to present an analytic treatment as case example for this book and refer to analyst and analysand, I will in the following paragraphs roughly outline the varying psychotherapeutic conducts resulting from the decision to choose either a psychodynamically based psychotherapy (psycho*therapeutic* conduct) or a analytic psychotherapy (psychoanalytic conduct).

Long-term analytic psychotherapy is developmentally characterized via a temporally concordant deepening of therapeutic alliance and working *in* transference: as if we sit with a child whose earlier experience indeed cannot be healed but not the less mourned in order to regain a future outlook on life (Orange, 2004, S. 50). The developmental therapeutic process occurs primarily within the consulting room.

Psychodynamic psychotherapy is primarily focused on conflict. Treatment aims at working *with* transference to engage in a developmental process by establishing a procedural 'focus' for interpreting pathologically active interactional constellations of conflict and maladaptive relationship patterns. The process takes place mainly in the patient's life outside the consulting room.

Psychoanalytic conduct guides the systematic reflection of interaction as it occurs in the interaction of relationship, transference and countertransference. This conduct emphasizes the metacommunication – the thinking about the thinking of the analyst and the analysand.

In contrast, psychotherapeutic conduct emphasizes the reflection of problematic relationship and interaction patterns in social reality of the analysand. Conflict resolution is actively stimulated by encouraging the analysand to act and learn actively through trial and error. Communication within the therapeutic session targets action outside the therapeutic situation.

Psychoanalytic conduct remains contemplative by encouraging taking a distance towards the social conflict in order to allow space for observation, thinking about and contemplation, therefore postponing taking action.

The psychotherapeutic conduct is symptom-orientated, goal-orientated, conflict solution-focussed and focally centred and therefore strongly structured. Psychoanalytic conduct is free from a prestructured approach to the development of an inner space. The freedom for development of the inner space remains protected from assessment and devaluation via a safe outer framework. Within psychoanalytic conduct we assume intrapsychic perspective and intersubjective unfolding of transference and countertransference rather than currently pressing needs for conflict resolution in social reality *out there*.

While psychoanalytic conduct targets a change through reflection of the interaction, psychotherapeutic conduct strives for change through interaction itself (Hohage, 2001).

The analyst's competence in either of the above means of conduct consists in the ability to ponder the therapeutic process of choice with methodological certitude and readiness for investing into the relationship while knowing that the reality of each treatment session calls for psychodynamic interpretation and amplifying questioning itself leading to the *Analytic Third*, creating encounter, change and individuation.

Preliminary sessions with the analysand in my case example led us to assume the possibility of a crisis intervention focusing on the current fears attributed to his intimate relationship crisis. However, these fears turned out to have been recurrent throughout his life and within all relationships, impeding deep and satisfactory relationship bonds. Deeply stipulated inner attitudes and inhibitions existed, preventing a personal maturation and development, consequently also obstructing a satisfactory and successful professional life.

Out of his existential crisis of meaning and development as well as the potential possibilities to understand them psychodynamically, both concurred a clear indication for an analytic psychotherapy.

Therapeutic goals and work alliance

We agreed on meeting three hours per week. Mr. M. wanted to sit first: 'I can't imagine losing sight of you.' He was afraid of not being able to speak while lying down and found sitting in an armchair more comfortable. When I said: '... and maybe you don't want to give yourself up to me like that either?' he smiled and hesitantly owned up to the idea. I offered he could try 'the other situation' later, when he felt safer with me. After months and an *apocalyptic* vacation with his girlfriend, he lay down on the couch: *'Today it has become clear to me that I am tired of the confrontation with you vis à vis, something has to move now.'*

During important times Mr. M. brought drawings and pictures to sessions. I asked him, both out of respect for his originals and to be able to think about them further, whether I could make a copy of them, which he generously allowed me to do.

My personal initial treatment considerations were predicated on an obsessive-narcissistic neurotic problem and the focal problem of the neurotic 'eternal youngling's' individuation, taking responsibility for himself and others, working reliably and finding one's own creativity (Franz, 1987, p. 58ff.).

Shortly before the end of the therapy, he proactively sat facing me again for the last few hours. Using the following words, he brought the therapeutic process to a termination before the full scope of the approved hours was completed: *'After all, there are sicker people than me.'*

Chapter 8

Context-guided treatment practice

The individualized treatment method of analytical psychology

If we presuppose that every analysand holds a preparedly available readiness to all possible forms of his psychological development, whose development rests not least on the hopes of the analytical relationship, concludingly the analyst must adapt to the analysand's developmental requirements with the greatest attention and intuition (Lesmeister, 2005, p. 266ff.). For this it is imperative to pay close attention to the analysand's previous life and relationship experience in order to let the panorama of his biography speak to the analyst.

Starting point is then the experience of the analysand, his own 'story' and its emerging effect in his current way of life.

The analyst's inner demeanour should include letting the individuality of the analysand live within himself, granting the analysand an inner space and allowing him to leave an *imprint of his being* (Lesmeister) there. In order to find access to the aforementioned developmental possibilities of his Self, the Self of the analysand needs to be understood empathetically and from within through a process of adaptation or 'mimesis'.

Grounded in these expectations to the analytic demeanour, there is a compelling demand for a high degree of individualization of analytical psychotherapy. Opposing ideas of a 'standard technique' adhering to what could be 'right' or 'wrong' in treatment practice, it ought to be guided by the former life and development history of the analysand and orientated towards the fulfilment of his individual needs. The 'Art of Treatment' to which Jungian psychoanalysis is striving for consists in capturing as many external and internal conditions as possible, ideally comprising the entire 'life context' of the analysand.

Furthermore, we need to account for the reflection and influence of the analyst's life context in the analytical process. As analysts, we want to understand what the world is like for the analysand – and we also want to influence his world by inviting him to participate in our thinking, our world and our perceptions. Additionally, analysts stay alert to the dynamic of the declarative, autobiographical memory system, which is continuously reworked. Within the autobiographical narrative of the analysand, we highlight his achievements and abilities as well his positive developments despite various potential limitations.

In our present treatment example understanding unreservedly how the analysand's radical disruptions in his self-confidence had occurred from his biography's many insults and injuries was elementary. Those disruptions included: to be an unwanted child, numerous situations of loneliness, the inaccessibility and violence of the father, a betrayal of the mother, father's turning away from him after the birth of the brother, to be actively left by the parents to the violence of a teacher, the shame, to have to wear inappropriate girls' clothes without parental support, bullying of classmates, school failure, parentification or attachment by the mother after the father's death, sexual fears towards women, overstraining oneself in the university thesis and many more.

These events portray the analysand's life story as a long sequence of insults, failures, defeats, setbacks and fears. In repeated confrontation he tried to lean up against them and protect himself from them, did not cease trying to assert himself against them and has even been led to the possibility of putting an end to his life as a last revolt against them.

Other biographical lines worth mentioning are his lifelong struggle with a transgenerational 'shadow of power and cruelty' and a lifelong examination of his artistic talent, which led to an interesting and satisfying professional activity and perspective. Women have accompanied and supported him time and again but only very rarely a short-term friend. The brothers have always been strangers to him.

His relationships were mainly symbiotic: those who were close to him or came close to him became like 'body parts', all too familiar and all too threatening at the same time. The struggle to see his girlfriend as a partner with a life of her own accompanied the analytical process to the very end. Only when it became possible for him to dance flamenco with his girlfriend as a partner, to enjoy an old musical enthusiasm together as a dancing couple, change became apparent. Moreover, he became able to perceive himself as part of the familial triangle – through experiencing the peculiarity and individuality of his (eventually welcomed) son, as well as the curious embracing of his fatherhood, leading to a desire to *wanting-to-be* a father to this child and the acceptance of the common parenthood of the couple. In this, mother, father and child are connected and can simultaneously locate themselves in their respective places. Thus, a secure differentiation of self and others had become possible.

Spoken dialogue – body language – performative dialogue – scene

Articulated in his 'dialectic method', Carl Gustav Jung put forward a dialogical principle. This principle currently experiences a revival particularly within philosophical and developmental psychological fields of intersubjectivity theory, sustaining its continued development. At the heart of its principle rests the constituent meaning of the 'unknown other' in the development of the Self.

This dialogical principle changes the entire psychoanalytic process as well as the understanding and handling of transference and countertransference. Furthermore, it offers new perspectives for the problematic themes of abstinence and neutrality. This dialogical principle grants an increased symmetry and reciprocity in the analytic-therapeutic relationship, which in turn becomes more receptible for new content and meaning during 'now moments' or 'moments of meeting', which lay 'beyond interpretation' (Stern et al., 2002).

Within our case example psychotherapy converged into a dialogical process, as soon as the reciprocity, empathy and responsibility unfolded and was experienced by the analysand. He became aware of a 'common job', which was shaped in the meeting taking place. Those moments of meeting were particularly fertile when we both explicitly or implicitly shared consensus of entering a meaningful access to conflicts or an emotion that has been initially incomprehensive.

Frequently a performative dialogue, spoken through bodily signals, unveiled access to such aforementioned meeting moments. For example, the analysand informed me of his current mental state via his mimics, physical posture or the way he shook my hand. When his state was bad, I immediately felt gripped by a feeling of concern and I observed myself calling him overly courteously into my practice. With time I also learned to pay attention to his expressive body tension and other physical reactions, such as sweating or intestinal noises, which particularly pointed to current preconscious conflicts placed amidst our current verbal dialogue and referring to us as well as his own physical counter-transference reactions. In the process, it became possible to recognize the symbolic value of such events and to make comments such as: '... obviously someone else has something to say', which, through the humour associated with it, often opened up a more relaxed approach to the respective topic.

Another important physical indicator were his headaches. They regularly coerced him to withdraw from people and situations. For a long time, the unconscious exit routes out of conflictual constellations offered by those headaches remained opaque. Increasing familiarity with its functional undercurrents decreased the prevalence of its symptoms. However, if headaches did arise, they caused all the more reason to pay an even closer look at their indicative leads.

Additionally, the development of an increasingly shared scenic understanding has been specifically powerful in the clarification of relationship conflicts to his girlfriend. The following suggestions could open amplificatory ideas as well as trigger memories of conflictual situation within his family: '... if we put as a theatrical scene what you have experienced at home yesterday – does something appear familiar to you?'

Symbolic attitude and rêverie

The process of symbol formation is anchored in intersubjective experiences. Trauma can impair the symbolization function itself or the adequate use of the symbolization function, i.e. the *transcendent function*. Examples found in the analysand's case comprise traumatic blockages that can be related to the early violence experienced in his education, to the father's turning away from him and to the father's death in his presence, after which he felt himself a 'murderer' of the father, but also to his angry disappointment, which led him to knock over and put into a ditch the pram bedding of his newborn brother.

For the most part, the analytical process with him could also be described as an examination of the symbolic value of his dream images, fantasies and pictorial representations. Initially, the demeanour of the analysand to handle symbols he created from the analytical situation was rather flippant and disrespectful. Again and again I tried to draw his attention to his dreams and images, which often remained present and presented themselves to me after the sessions, often distracting me from other intellectual work by way of my working with these inner images as "work as double" ('Doppelgänger'; Botella & Botella, 2005). For example, the analysand's image of him crawling alone, injured and deaf on the street, and collecting broken glass (see the 'Burdens from biography' section in Chapter 7) revived images of the fears and suffering of a prolonged stay in hospital in my own childhood. After I told him: 'I can imagine what it is like to become deaf and numb again and again', it became possible to speak very precisely about the different qualities of his experienced suffering from during periods of loneliness. What he had previously experienced as emptiness and times he couldn't recall in memory now became visible as an unrepresented traumatic gap in his experience. Through this, discovery of his anger at this fate conceded granting it to be a shared meaning of trauma and at the same time as an impulse for development (Maier, 2014, p. 369ff.).

Chapter 9

The midwifery method of analytical psychology

In classical Freudian psychoanalysis, it was the task of the analyst to search for repressed memories and (drive) conflicts, much like a detective or an archaeologist, in order to make the analysand aware of them. The conflicts of life with which the analysand was currently confronted in his everyday life seemed to be of psychodynamic meaning only to the extent by which they expressed or represented earlier, repressed and internalized conflicts. The primary task of the psychoanalytic process therein consists of revealing those conflicts (Arbeitskreis OPD, 2009, p. 112).

In comparison, C. G. Jung assumed early in his work that the memories of the past always become constructed in the present. All effective factors leading to regressive neurotic introversion can be deducted from the current conflict and in the here and now of a symptomatology (Jung, 1913, § 373). At the same time, Jung saw possible solutions for the respective conflict constellation as are readily available in general form via the contents of the analysand's shared/collective unconscious, inasmuch as each individual conflict simultaneously corresponds to a generally human, often timeless conflict constellation.

Jungian analysts believe that this general or archetypical template provides individual pathways to solution. Dream images, dream symbols and other imaginative and creative impulses are actively provided by the analysand's creative unconscious processes to engage specific avenues of solution and provide corresponding conversion energy. Jungian analysis therefore seeks to find a shared entrance door to these symbolic contents and their libidinal power.

As already expressed above, the methodological attitude of the Jungian analyst is oriented towards the Socratic 'art of midwifery' (Maieutiké): by way of amplifying references and questions, the analysand himself should find access to those contents of his own unconscious that are relevant for his current obstacles to individuation rather than through defining interpretations.

Structure and functions of the Self, conflicts of the ego

In analytical psychology the concept of 'Self' refers to a person's totality comprising all unconscious and conscious levels of one's mental and physical life as 'embodied Self'. The inner archetypal image of the Self often corresponds to the personal conception of God. It is frequently symbolized as sun, circle, in a quadruple form or as mandala. The 'Self' of the analysand in our case example symbolized itself via an inner image appearing to him several times in situations of great despair: *'Inside me is a warm, shining sphere. It's indestructible.'*

Children are born with an unmistakable identity and presumed 'primary Self' (Fordham, 1985, p. 90/91), providing fertile ground for many development potentials. This 'primary Self' turns to emotionally significant others with an expectation of relationship ('deintegration'). Experiences are reintegrated into the child's 'Self', granting its increasing abilities to live out its centering, organizing and relationship-creating function. 'Conjunctio' is what Jungians call the ability to enter into a transformative effective connection with another person, both internally and externally. The extent to which experiences can be reintegrated into the Self depends on the quality of the encounter and the ability for inner connection/'conjunctio' with the other person.

Apart from the archetypal dimension of the Self and its connection with the shared unconscious, outcomes can also be functionally captured via ego functions and relation building capacities.

Important *functional* aspects of the 'Self' that are accessible to consciousness have been described by the 'Operationalized Psychodynamic Diagnostics, OPD' (Arbeitskreis OPD, 2009, p. 117ff.). It lays out the functionally accessible structure of the Self in the following four dimensions:

- Self-perception and object perception,
- Control of Self and relationships,
- Emotional internal and external communication,
- Inner attachment and outer relationship.

The neurotic and long-lasting dimensions of conflict experienced by the ego in interaction with others are defined by OPD as follows:

- Individuation versus dependence,
- Submission versus Control,
- Reliance versus autarky,
- Self-worth conflicts,
- Conflicts of guilt,
- Oedipal conflicts,
- Identity conflicts.

At the beginning, the analysand had great difficulties in perceiving others in a holistic and realistic light and to negotiate his relationships with self-confidence. His self-esteem fluctuated. It was particularly noticeable that he hardly had any positive inner images of a couple and a couple relationship.

This is of great importance in that the different aspects of the Self and the different *self functions*, whether projected onto others or experienced inwardly, are represented in unconscious fantasy by a series of pairs placing 'Self' and 'Other' in a binary relation. Such images of inner couples would be, for example: the nursing mother and the child, parental relationship to each other, an inspiring couple like Shiva and Shakti, being embraced in care as Container-Contained, Mary and the child Jesus, analysand and analyst, vagina and penis, self and old sage, Philemon and Baucis. Pursuing, the regulatory power of the 'Self' aids the integration of the various, connected yet often conflicting, inner pairs. Consequently, the experience of an interconnected and enduring inner pair ensures the psychic survival of the 'Self' (McFarland Solomon, 1997; Bovensiepen, 2009; 2019, p. 98 ff.).

If negative or 'deficient couple' attachments dominate the 'couple in the unconscious', the 'Self' can make use of premature forms of defense, such as dissociation, and pathological forms of projective identification in order to render these pathogenic sub-complexes harmless.

The 'deficient couple' in the analysand's unconscious triggered strong fears of individuation and dependence and of being influenced and controlled by others.

The structural characteristics of the analysand, i.e. his impairments in all four dimensions of his self-functions, had an enormous influence on how he defined himself and the type of conflicts that he could experience and overcome. At the beginning he had only limited access to the affective and the distinguishing and differentiating functions necessary for the regulation of the Self and its relationships (Rudolph, 2005, p. 48ff.).

Shaping the dialogue with him so that his *inner working models* of relationships and emotional patterns of expectations could develop to *mature relationship functions* became an important part during the course of treatment.

I would like to explain this using the example of the analysand's development of sexual experience. At the beginning, his sexual behavior and sexual fantasy were rather self-centered and very easily disturbed by inner images acquired in lifelong patterns of dealing with pleasure and satisfaction. With growing confidence and inspired by a dream in which I fell in love with him (see the 'Dream series and the analytical relationship' section in Chapter 10), a transition space opened up in which repressed sexual desires were permitted and could be addressed (Quindeau, 2014, p. 117).

Inner-psychic structural characteristics are above all anchored in implicit relationship memory, so that the microprocesses of our emotional attunement and its restoration paralleled a very important quasi-musical fine-tuning of us relating to each other.

Transference, countertransference, resistance

From the initial interview onwards I liked the analysand because of his un-conditionality and the hope of still being able to escape the dead ends of his previous life. During the treatment process, suicidal impulses occurred that were so worrying that in one case I had to break the analytical framework by investigating whether the analysand was still alive.

A phase of the sensible flattening of the analytical process, which was void of dreams rendering, was difficult to bear until it became clear to both of us what might his then-persistent, rationalizing resistance had. At least three sources of resistance became clear: the shame of talking about certain fantasies, the fear of moral condemnation by the analyst and the need for individuation by wanting to be self-sufficient, independent and 'adult'. A certain impatience on my part as a form of countertransference strength-ened the patient's resistance until he dared to complain about it.

The most intense experience in the course of my countertransference was the individuation process of the analysand. In the last third of the analysis I was increasingly confronted with a very serious, mature man who, accom-panied by severe crises and accidents, had discarded everything that had made him 'puppet-like'.

Shadow recognition and shadow integration

Mr. M. kept coming back to the tragic fate of Jimi Hendrix, whose mu-sic fascinated him. To him Hendrix stands for the ecstatic, which Mr. M. felt strongly attracted to, since eating, drinking, consuming, loving but also working were all possible for him only in 'ecstatic seizures'. Hendrix also stands for the closeness to death, for ending life.

With time we were able to classify his often-strong suicidal impulses as a backdoor, or last resort, in order not to having to face up to changes and responsibility. He recounted buying books on the 'history of death' and one with 'farewell letters from suicide murderers': *'Perhaps I want to understand the finiteness of life or get to know life in its dimensions.'* He thought life after death was impossible.

One day, shocked, he noticed that for a long time he had looked at and treated family members, friends and acquaintances like parts of his own body. He was very frightened by this form of incorporating other people: *'This is completely absurd for me today, I want a counterpart with whom I can negotiate.'*

In the process of replacing the idea of a love relationship as merging with a partner, one day he fantasized about the rape of a female burglar or im-agined falling in love, making the woman a child and then leaving her. Two aspects went into the imagination of 'child making': the male identification with the 'father hero' who begets – and to sneak into a woman's love through the womb – as a child. Or he imagined himself becoming unfaithful to his girlfriend on the occasion of an encounter with an attractive woman. He thought of himself as being really creative only when 'in love'. He needs an inspiration like a painter needs a muse. He believed he had to lift the anchor, the balloon had to fly freely (see Figure 10.1 on p. 102). A fierce Anima obsession had gripped him in the form of a flash-like projection of erotic fantasy: 'Like an avalanche, unable to stop.' At the same time, he could fully develop his creativity. For the first time, he experienced himself as a 'whole man'; everything seemed possible and easy to do. The groundbreaking amplification resulted from a film that impressed him, namely *Damage* by Louis Malle. This film is about an overwhelming and destructive love (*amour fou*). Mr. M. found access to his own destructive impulses towards his girlfriend and subsequently could turn his attention to her again.

Emergence of transformation

The relationship experiences in the analytical field have enormous significance with regard to the question of how transformation arises from an analytical process. In 'moving along' the analytical encounter, sudden events occur that the Boston Process of Change Study Group has termed as 'Now Moments' or 'Moments of Encounter'. In such moments of encounter changes take place through sometimes minimal, sometimes dramatically shattering non-verbal and affective interactions, which both alter and anchor themselves in implicit relationship memory (Mertens, 2015, p. 189ff.).

In any effective psychotherapy, 'numinosis or magical moments' and situations occur unintended by the participants, leading to an intensive experience and increased correlation of the analytical couple. Such moments have also been described as experiences of the Self, as the unthought, as synchronistic events or as 'eureka moments'. All of the above comprise processes of abrupt consciousness of contexts and meanings that suddenly expand the inner horizon. They can result from touching dream images or an analysand's meaningful thought and equally from an interpretation, amplification or action by the analyst. Such moments can pose the starting point for dramatic changes of direction in psychotherapy, often marking turning points in the analysand's life.

Pursuing, I will outline an example of the descent of particularly touching events and sudden insights upon the last third of the analysis with Mr. M.

Due to a difficult domestic situation with his partner and the screaming and teething baby, he had grown increasingly quiet, sad and eventually acutely suicidal. He didn't show up for the next session, leaving me waiting without cancellation or message. I was very worried as I could not reach him by phone. With growing concern, I decided to seek contact to one of his friends, whose job he had once mentioned. I finally found this friend and reached him by phone, who in turn drove to Mr. M. and found him at home. Subsequently, Mr. M. called me and I gave him a same-day appointment.

Mr. M. reported in a very agitated state: he was completely beside himself. But he had decided once and for all not to lay a hand on himself now and in the future: *'I realized that so far I have only lived for women and have been invigorated by them. Again, and again I looked for the mother in women.'*

My efforts to find him had surprised him very much: *'... I would not have expected you to put so much effort into it for me'*. For the first time he consciously perceived my relation to him: *'You are really involved; I always wanted to see you as neutral.'*

Our relationship had suddenly changed and deepened enormously. This change was triggered by the analysand's emotionally tangible insight into the nature of his relationship with women and his girlfriend and the perception of my concern and relationship with him. We both felt energized by this tangible contextual meaning of his insight and his life, which could previously only be addressed 'lifelessly'.

It is not presumptuous to speak of gracious occurrences about such moments, when thoughts can suddenly be thought of in terms of vivid meaning. In the example above, the decisive thing happened at an absolute suicidal low of the analysand's life, accompanied by an absolute culmination of the analyst's fear for him and his life.

The hours after were completely different: calm, intense, profound and relaxed. Everything felt like after a violent thunderstorm, when the wind dies down and the sun breaks through the clouds again. The portrayed situation can also be described as the occurrence of an intense experience of both with one's respective 'Self' in the analytical field and as a climax in the analytical relationship. Frequently, such moments of encounter occur when analytic progress stalls and this stagnation must be temporarily interrupted, as for example via the analyst's action/enactment laid out above (Crowther & Schmidt, 2013).

Chapter 10

Consciousness of the unconscious

Key coordinates of dream work in analytical psychology

At the core of our personality lies our ability to dream. This ability recognizes the fleeting and poetic dream mode of our psyche as well as the intuitive-symbolic 'thinking' of our Self (see 'The Intersubjectivity of dreams in the analytical field' section; Braun, 2010b).

Psychodynamic psychotherapy and psychoanalysis are primarily concerned with the transformation of the psychological effects of problematic current and past experiences of relationships. These denote inappropriate actions, negative relationship expectations, paranoid and hostile attitudes and perceptions. Such harmful attitudes can be recognized and questioned through the interpretive work with dreams (Kast, 2006, 2010).

Looking back on the past, but also for the here and now and in view of future developments (finality), the dream can not only assume a balancing 'Orientation Function' with regard to the past and the present but also mark future developments (finality) by highlighting pathways towards solutions for various conflicts, particularly those in transference.

Particulars of how the analysand deals with and remembers his dreams, specific times when he brings them in and how he communicates them are highly specific to the respective relationship constellation. Thus, the context of transference plays a major role in the experiencing of dreams. Experiencing comprises what dreams are dreamed of, which dreams are remembered and told and how they can be interpreted in a consensual manner. Transference is an unconscious organizing principle of the psychoanalytic process, which also assigns its 'place' to experiencing dreams. The organizing principle of transference regulates affects, influences the type of symbolic representation and has a reorganizing function (Deserno, 1999a). Furthermore, transference is a transitional phenomenon since it mediates the relationship situation outside and within psychotherapeutic treatment.

The intersubjective aspect in the practical work with dreams must be taken into account. This is particularly the case in the analyst's own countertransference. The interpretive work of the analyst and the analysand on the respective dream therefore integrates all three: the perspective of the investigation of 'transference in the dream', the emergence of the dream 'in the transference' as well as the perspective of the analyst's countertransference notions (Deserno, 1992, p. 975).

I assume that everything that happens in a psychoanalytic hour is related to the dreams that an analysand reports in the respective hour (Dieckmann, 1983, p. 58). In a certain sense, the dreams are dreamed 'between' the analysand and the analyst (Jung, 1981, Letters I, Letter to Dr. James Kirsch of September 29, 1934, p. 223). The dreams arise not from a specific characteristic that is a priori important but from the bi-personal intersubjective field of analysand-analyst. Within this field that specific characteristic becomes significant for their emergence as analysand/analyst. Like figure-ground, dreams are the result of a joint creative mental-emotional process (Moser, 2003, p. 745).

The event of the dream represents processes of symbolization and the generation of new meanings that occur during treatment and that emerge from the encounter between the analysand and the analyst (Ferro, 2003, p. 136ff.). We must be able to perceive the dream narrative as an event of relationship and let ourselves be touched by the analysand's dream (Kreuzer-Haustein, 2000) ergo accept the dream as a 'gift in transference' (Morgenthaler, 1986).

In the course of an analysis, dreams can act as 'transitional objects', fed with meaning conceived from the relationship constellation. Analogous to an infant who, in a certain phase of development, chooses a cuddly toy or its blanket as a substitute for the absent mother and endows it with motherly 'qualities', certain dreams can be remembered and interpreted alternately in the absence of the analyst. This function was found in the initiation dream of my analysand (session 28, see p. 97). Depending on the state of the therapeutic relationship, it served either as a comforting sleep fantasy or as confirmation of the analysand's enormous fears of being abandoned that were grounded in past experiences. In due course, the dream has lost these functions when the analysand could take on the role of being a father to his own son.

Single dream and dream series

A single dream can deeply move us and capture our mind. But it is especially the observation of dream series that gives us deeper insight into unconscious conflict constellations, pathogenic complex parts and important parts of ego-structural characteristics of the analysand (Jung, 1931, GW 16, § 322).

The dream series of a given therapeutic treatment includes all dreams that are dreamt during times for preparing for treatment and the course of its duration. It describes and comments on the temporal course of certain structural changes and transformations in the communication of the analytical pair. The development of the analytical couple is the essential context of the dreams communicated during analysis (Jung, 1943, GW 12, § 50). The dream series describes the individuation process of the dreamer in the intersubjective field of psychotherapeutic treatment. We call the first dream communicated *initial dream* and the first dream with emotionally touching archetypal images as *initiation dream*. Initial dreams often show briefly in symbolic form the main problem or main feeling of the analysand at the beginning of psychotherapy. They can also refer to the possible and unconsciously intended course. *Initiation dreams* often stimulate change processes.

A major obstacle to individuation processes is the unconscious attachment to the imagines of damaged early relationship persons and the early pathological interactions with them stored in the affective relationship memory (Lambert, 2002, p. 198ff.). This leads to disturbances in the processes of mentalization and in the development of attachment capacity. The affinity to those imagines forms the emotional background on which the 'micro-world' of dreams is projected. The hypothesis of the intersubjective composition of the dreams communicated in the analysis allows the dream figures acting in relation to the dream ego to also be understood as imagines of the analyst experienced in the transference.

In our treatment example, the dreams repeatedly had decisive hint and transformation functions. Over and over again the essential problems and conflicts were presented in the entire dream series, brought into impressive, often dramatic pictures and scenic sequences, 'played through' with great emotional participation and were further developed. In dream symbolism, numerous obstacles to the individualization process of the analysand became apparent.

Intersubjectivity of dreams in the analytical field

The analysand's dreams illustrated predominantly interpersonal themes of guilt and fear.

They can be found in the course of treatment as conflicts from the life story transferred to me and as newly arising conflicts from the therapeutic relationship. They resulted from my reactions and from the progress of the analysand, which newly found access to previously blocked aggressive impulses. At the beginning of the treatment, conflicts 'in' the transference (Körner, 2014) appeared in his dream material, e.g. its burden and inhibition of development, the issue of guilt and being condemned. Later, themes developed 'from', out of the transference situation experienced, such as love or hate, being robbed, disappointment.

In the following I present the interpersonal constellations of conflicts depicted in the analysand's dream series, omitting other dreams and dream elements. The dream material is given as block quotes in this chapter, dots indicate omissions. The psychotherapeutic session is indicated in which the dream was reported.

Dream series and the analytical relationship

Mr. M. came to the first hour: *'Here I am.'* He seemed a bit confused but more animated and excited. I too was curious. He reported feelings closer to real-life events since the preliminary talks. Following, he reported a dream, his initial dream (session 1):

> I broke up with my first girlfriend in a long and very strenuous fight. It was a reckoning.

The heavy accusations against him that he could not understand her had been exhausting and emotionally very moving in this dream. He suspects that the dream might be about separating from a former girlfriend; that's all he could think of.

A central theme of the analysand that there is insufficient separation from his 'first girlfriend', his mother, was clearly and concisely named in that *initial dream*. Taking a prognostic view, I was glad he successfully performed the detachment in the dream, even if it was with great efforts.

Instead of interpreting, I followed his line of thoughts: the separation from a former partner, which was not 'properly' carried out, would still affect his relationship with his girlfriend today.

His ecstatic infatuation quickly faded away after his girlfriend got pregnant and then lost the child. He was deeply shocked at the idea that she could now chop a nail in the 'sacred walls' of his apartment without his consent, put up any furniture or even hold a housewarming party: *'I have nothing to celebrate.'* Strikingly, his talk about her moving in was often accompanied by strong stomach growling: *'I'm afraid of symbiosis. Where am I as a man? If a child were born, I'd miss out.'* I drew attention to the growling stomach, which in the future came up even more often when it was about a childlike need to wanting to feel safe and well-fed. He remembered pictures of his childhood: *'I stand there like King Louis, hands on my hips, happy.'* After the arrival of his first brother and his father's turning his back on him, the picture had changed: *'I often felt insulted. My new nickname was:* Mimosa.*'* From then on, he always got the short end of the stick. Even at times when he was physically and mentally stronger, he was eventually unable to execute the decisive blow.

He became very sad and reported the following dream (session 20):

> I find myself at a party with many guests in my new apartment. I get rest-less among these people after a while. Individual guests, especially foreign guests, smoke. I look for a suitable spot and then I shouted: 'No smoking in my room!' As if inevitable, people stop smoking. At some point I am alone in the room and begin to paint a wall white, I do not really make any progress, I despair that I have to work with a kind of 'brush replacement'. The paint drips on the carpet; I scream: 'Why won't anyone help!'

Ideas of the analysand: *'I never considered that the girlfriend would bring her own culture. I'm afraid to give myself to it. I didn't think the foreigners would stop smoking; they can be so impertinent. I want to renew something with this color: white as the basis for a new beginning.'*

At first glance the dream referred to a real problem: his girlfriend had asked him for a move-in party. The current conflict was his fear of drowning in his partner's circle of acquaintances. With a view to the future, however, the dream ego succeeded in making himself heard and respected.

On the subject-level it was about the 'inner man' and the ego complex, which must gain acceptance among the other complexes and provide 'pure air', spiritual clarification. The hard work is the 'whitening' of the wall. This is about creating order in the chaos of the past – and perhaps also about the 'whitewashing' of repressed conflicts, as the 'dripping brush replacement' made one think of erotic difficulties. The dream ego is helpless, lonely and desperate. There is no visible relationship with the analyst. Personal and strange impulses pursue into the analytical space and into the developing Self of the analysand – into the 'new apartment'. In the transference Mr. M. felt left alone by me: I had not contradicted the move-in plan of the girlfriend.

With her moving in, he experienced a slow change. He began to accept her in his studio flat, while also being confronted with his strong compulsiveness: she would use his things, would cause *'dirt'*. This has disturbed him before in the family: *'All the dirty things flow into each other – sex no longer plays a role, is pushed away. Father was the only one who could draw boundaries.'* He in turn was not able to do this; he had previously only been able to shield himself through stubbornness.

The second part of the treatment was initiated by a very moving dream, the initiation dream. This one described his being stuck in a situation and a temporary regression. He was in such a way submitted to the forbidding mother imago that he looked for an extra flat for the girlfriend and forced her to move out. In this phase he was internally and to some extent practically inventing numerous personal and professional 'solutions', each of which soon lost their attraction again.

He dreamt (session 28):

> I'm on a construction site in a wooded area, working there. A large concrete bridge or runway is being built, a plateau standing on pillars. Until the concrete has dried and strengthened, the plateau must be supported. I had previously already worked as a support pillar on this construction site; it was very hard to bear. The pressure increases immeasurably when the concrete is poured into the formwork. I tried to find a way out, a technical solution. I was provided with a good-natured man, heavy built like a weight-lifter. We tried different positions; how we could bear it together: but he would be hurt by co-supporting. I had an idea: 'why

not take a beam or a log for support?' I presented this idea to the engineering office, but the entrepreneur's wife did not want to allow my fate to change.

He experienced this dream of being a pillar of support as very strenuous. He had woken up in the morning with heavy neck pains. The following ideas came up: the entrepreneur's wife is the sum of all mothers and women who want to place me in structures: *'I am then at their mercy, have to stand upright, that is no longer dependent on me. My mother is always pestering me with marriage. Why in my dream don't I have the idea of just leaving? I have gotten into a special position in all my jobs. It takes forever for the concrete to harden ...'*. I recalled the mythological *Atlas* to him, who carries the world on his neck – Mr. M. could not make any use of this amplification: 'carrying the world'- that was too much for him. He remembered that he has long wanted 'to write about people, who by looking at them you can tell what they do: asphalt workers, slaughterers'.

In sum, this dream was about working reality on the one hand and about the creation of a landing strip in the 'wooded area' of the unconscious on the other hand.

In different sessions during this part of treatment we came back again and again to the dream of being a pillar and to the helpless position of the dream ego, thus helping him to recognize his present and past relationship constellations.

Behind the dream image of the refusal of the 'technical solution' by the entrepreneur's wife rests her loyalty to an 'entrepreneur' – the father – and in this image he recalled his submissiveness towards his mother.

The fact that the weightlifter's help was not enough was a clear indication of his uncertainty about my resilience and suitability as a 'soul helper'.

The time following this dream was thematically about his professional life and the insufferableness of it. It was about the impulse to emigrate and to make one's fortune abroad. Other topics were the envy towards his well-qualified girlfriend and the fantasy that someone would take him by the hand and say, 'That's the way!' Again and again and with strong ambivalent feelings, his thoughts revolved around the lack of experiencing his father as a counterpart. At that time, the analyst was still felt as a 'bulldozer', whose 'towing aid' could shake the analysand's defensive structures ('embankment') far too violently.

He was dreaming (session 36):

> I drive a car ... it's a dead end. I get stuck turning the car around. A bulldozer dragged me off: then the whole embankment of the road broke down. I woke up loudly protesting.

The developing closeness and intensity in the analytical process caused strong fear of or aversion to homosexual seduction by the analyst.

He was dreaming (session 62):

> I'm in a therapy session. You say you fell in love with me. You were
> gonna take my hand. I said no, I don't want to! In my dream I asked
> myself: What do I continue to do? Should I stop the therapy? Then I'll
> tell you: that's your problem!

He could accept my interpretation that the same sex transmission might
include the desire for a friendship alliance, for a combat companion (Jung,
1974, p. 209).

A few weeks after the decision to move out, the girlfriend told him once
again that she wanted a child from him. At first he was very depressed, did
not feel up to that task again and rejected the thought of it.

He was dreaming (session 71):

> I am in a comprehensive, never-ending test. It was about the ability to
> handle a marionette puppet, to perform very complicated moves with
> a whole bundle of threads. A tremendous stress, the doll moved very
> clumsy.

While he brought all the negative thoughts and feelings with him in the following hours, he got the impulse at home to draw a children's book (see 'The
integration of creative expressions' section; Figures 10.5 and 10.6, p. 107).

A shadow aspect manifested in him, now frequently reporting fantasies
about fights, also about *shooting someone*, which frightened him very much,
as well as the fear that he could run over a child with his van: *'Maybe 'I'm
hoping for a catastrophe so I can stop all this.'*

Correspondingly a dream from session 79:

> I was supposed to be executed; I was already convicted. It was clear:
> there's no possibility to question that; everyone wants to see an exe-
> cution and to have someone to blame – I've been obedient to it. The
> execution's chairman treats me very carefully, but I know from the very
> beginning that the death penalty exists and that it affects my case. It's
> very serious. I have a feeling this scene is going to be happening in this
> event here.

In the process of *working through* the oedipal conflict of guilt, the father
appears, who died in an accident in the presence of the analysand. Looking
back, when his father was dying the then 16-year-old analysand experiences
himself 'paralyzed'. Subjectively, all his life he felt guilty for not being able
to get help in time, although the rescue of his father was objectively not
possible. The analyst appears in his dream as a 'condemning one', and in the
service of morality and the Law of the Father, execution is certain.

In the dreams of the analysand up to the 80th hour, the dream ego appears helpless, clumsy, oppressed by sexual inferiority complexes and homosexual impulses, confronted with unsolvable or most difficult tasks and subjected to endless trials. The inadequacy and guilt of the dream ego did not allow for any solution other than a final condemnation and execution (crisis during the climax of treatment).

It was the dream of the 80th session that initiated a double transformation: Substantial shadow elements become projected to the dream imagines of the analyst, leaving the analysand relieved. An inner counter-movement begins: the analysand no longer wants to be led to execution, but now lets the analyst become the perpetrator and 'robber of his cabbage' in two appearances (session 80):

> I was driving a car. At some point two men got in: a tall, coarse one and a smaller one. I had quickly accused them both internally of having the intention to attack me – but nothing did happen, absolutely nothing! The two of them became familiar, jokes were made, I felt understood … When my tension became unbearable, I drove to the nearest police station and accused the two of having ambushed me. In the meantime, they had started selling cabbages out of the trunk of my car! The journey continued then to their final destination. I thought it wasn't okay to pin a robbery charge on them. I had the impulse to retract my charges.

Even within this dream, the shadow projection could be recognized and taken back to some extent.

As a matter of defense, dreams were rarely shared between the 80th and the 173rd hour. However, it was precisely this defensive process, the stopping of the submissive communication of dream material, which was the starting point for the individuation process during this treatment: He started to distinguish himself from the 'collective' of other dream-narrating analysands without already being aware of it.

In the second third of the analysis, the analysand was able to overcome the intrusion of shadow content and went on an intensive search for solutions. He entered into a more committed relationship with his girlfriend. He began to educate himself. At the beginning of the last third of the treatment, the threshold of consciousness, or maybe only realization of mentioning transference-related dream content, was again overstepped. It was no longer possible to suppress the problem of the unconsciously compliant adaptation of the analysand to the analyst's imagined 'orders'.

He was dreaming (session 173):

> During the therapy sessions I get the task of playing a role. You want me to act like an Asian. I identify with this new person; at some point I am completely him. I'm sitting across from you. You say: 'Your identity is

known to the authorities, you don't have to pretend, you can stop play-ing. I am stunned because I think I played the part very well.'

This dream shows that up to then a *healing in transference* was still going on: the perfection of a role-playing game imposed on him by the analyst or the parental imagines.

In the dream of the 'Asian role play', the problem of individuation con-stellated itself for the last third of this analysis: the active and self-confident overcoming of the analysand's stance of submission and his striving for au-tonomy and relatedness. After panic-like escape impulses, he was able to accept the pregnancy of his girlfriend and later developed an intensive re-lationship with his son. He began to unfold creative abilities, which led to a demanding and satisfying professional activity.

The integration of creative expressions

Apart from amplification of the dream symbols, symbols of images and his impulses to design drawings set the course of treatment. The analysand often felt the need to freely visualize fantasies and dream images in color or black-and-white drawings (Vogel, 2008, p. 96f.). Imminently, from the first sessions he reported on the urgent need to create a large collage of men in a multitude of life situations, representing an image of the masculine and its development goals as man. At the same time, an inner image haunted him after his girlfriend had expressed the wish to move in with him: '*In my apartment, two balloons of different colors expand more and more, squeezing me, almost to the point of suffocation.*' He felt reminded of swollen breasts, overwhelmingly maternal, also of soap bubbles, which shimmer beauti-fully, but can easily dissolve into nothing. Amplifications of the sorcerer's apprentice, who had set in motion a process of too much good that could no longer be stopped by his own efforts, accompanied the confrontation with his female relationships amplifying over a longer period of time. The maternal-feminine sphere comprises aspects of containing, detaining: everything born of it remains under its influence. Thereby we can think of a kind of 'psychic gravity' as a tendency of the ego to return to the original unconscious state. The depression of the analysand, a 'libido loss' of con-sciousness, can be expressed energetically as a discharge of the ego com-plex by the 'gravitational force' of the archetype of the (re)devouring Great Mother (Neumann, 1974, p. 40ff.).

The leading first picture for the initial period of treatment was a bright-yellow captive balloon in front of a brilliant blue sky, anchored in a green meadow.

To him, the colors represent the creative elements; anchoring means safety but also maturity: '*I'm actually too old to keep going back and forth.*' The bal-loon feels like a backpack: '*warm, lush, leisurely*'. He appreciates the solid anchor: '*Like the solid calves that appeal to me in women – a firm point of view.*

But the balloon is also an egg.' I contributed: 'a symbol of the beginning?' He often has the feeling that he is stuck in an egg like that but that the shells are just as stable as cobwebs: *'The thin thread to the anchor could get even thicker, maybe like an umbilical cord?'*

He was proud of his drawing. I had the feeling that I understood more about his situation: since he himself was not to be seen in the picture he had to remain hidden 'in the egg'. Certainly, a pilot was nowhere to be seen. At that time being 'on a thin anchor rope floating freely in the wind' became our image for his present situation. After gifting his painting, he changed in the psychotherapy sessions. The hesitant, perhaps even suspicious, made way for what I experienced as more openness. He remembered that in emergency situations in his life the following inner image had repeatedly appeared: *'Inside me is a warm, shiny sphere. It's indestructible.'*

My uncommunicated amplificatory idea was: the Atman/the Self as the antipole to the world is 'smaller than small and larger than large'. At the time of Mr. M.'s greatest (external and internal) chaos, the symbol of the 'Golden Egg' appears as a self-symbol for a new beginning. As an archaic symbol, it also refers to the extent and depth of decompensation. In his painting of the 'Egg/Balloon' (see Figure 10.1), this symbol becomes an uplifting moment

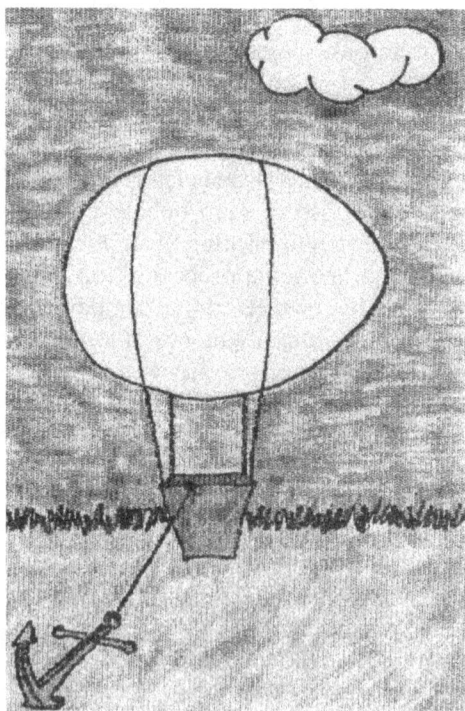

Figure 10.1 The anchored captive balloon.

that requires grounding and anchoring in order not to drift into the un-limited/cosmic. The necessary Anima or conversion energy towards both anchoring and uplifting was already 'on standby' in the girlfriend's desire for pregnancy but at the time still had to be fended off.

After some time, he could not only symbolically express chaos but also his everyday feeling: he had always felt like a black sheep, not like the others, as if not belonging to them, striving for other goals (see Figure 10.2).

Approximately half way through the treatment he brought along a postcard drawing: *'If I am well, I feel superior and strong, then I am on a raid without anyone noticing, I attach great importance to my incognito'* (see Figure 10.3).

The consecutive session Mr. M. arrived in a very bad mood: over the weekend his girlfriend had moved in with him. He felt 'as old as the hills' and that we would die soon: *'I came into a grinding mill. This is the worst case: I allow the entry, all the while beating my hands above my head.'* He had imagined fleeing to the Pyrenees, running into the mountains like Nietzsche's Zarathustra. A book he had been carrying with him forever. He remembered earlier retreats to his basement at home: *'my very own secret place'*. During that time, he had written short stories, remembering himself as very clear-sighted.

In this treatment phase, the analysand's image of a man, the 'Anthropos', rests symbolized in the narcissistic Zarathustra superhuman who, driven by fear, wants to walk the path into loneliness and mercilessness. Nonetheless,

Figure 10.2 The black sheep.

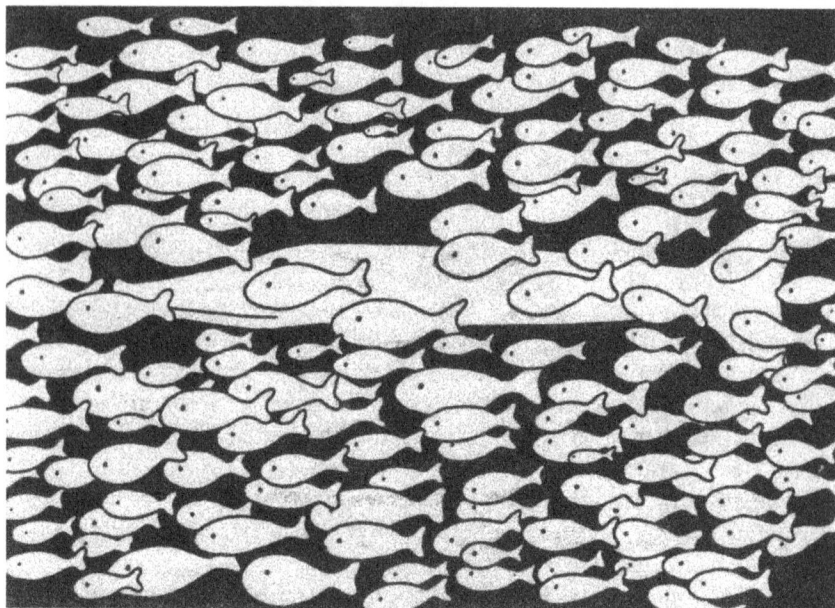

Figure 10.3 The pike.

Zarathustra symbolically represents an aspect of the unity and wholeness of his Self, albeit still hidden in the autocratic icon of a "vagabonding Christ" (Ribi, 1983, p. 137, p. 146).

Sometime later the analysand brought along the following day-dreamy fantasy:

> A cloud of steam rises from the water, floats above the water. On top of it is a castle. Slowly the picture floats away in the fog.

On meditating about his picture drawing, he suddenly and intuitively associated a backdrop: a giant bearded male figure carries cloud and castle in his hands, either guarding or creating. And this figure also has a backdrop: he emerges as a spirit from a bulbous bottle with a handle, which apparently had been uncorked.

About the castle he associates: a safe place, the place of his father, the Lord of the castle. I contributed that this safe place is also hard to reach, especially if there is a treasure or the Grail as in the history of Parsifal. He responds defensively: *'Now you're getting romantic!'* Me: 'Do you like "*Wolkenkuckucksheim*" better?' He: *'I've always liked to build them – but it's in the nature of things: Nobody can get there.'*

We know ghosts, or *Ifrits*, which were locked in a sealed bottle as punishment from oriental mythology. The tales of 1001 nights recounts the 'Story of the Fisherman with the Bottle' (Weil, o. J., p. 46ff.) and also in Grimm's fairy tale 'The Spirit in the Glass' this complex symbolic being appears, simultaneously dangerous and fortunate (see Figure 10.4).

The fairy tale goes like this: A poor day labourer had let his son study until he could no longer afford the tuition fees. Back with the father, the son finds a glass bottle with a frog-like creature within while chopping wood under oak roots. The creature imploringly asks for liberation. A 'guy of horrible size' appears out of the bottle and announces to the student that he is the 'great ghost Mercurius', enclosed in the bottle as punishment. He would now have to break the student's neck. Arguing well, the student manages to lure the ghost back into the bottle. Before his second liberation, he negotiates nonviolence and a miracle plaster, with which he can heal everything, as reward. The fairy tale concludes with the words: '... and (he) was the world's most famous Doctor' (Grimm & Grimm, 1985, Part Two, p. 53ff.).

Amplifying the other fairy tale of the fisherman and his demon can be understood as the necessity of releasing the spirit, or Ifrit, by the fisherman, since the traditional model of consciousness is no longer upholds. The

Figure 10.4 The ghost emerged from the bottle.

spirit symbolizes the 'revolutionary', perhaps destructive power of his unconscious, which rebels against the dominant paternal order (Dieckmann, 1984, p. 36ff.).

With Mr. M., the spirit takes on the form of an old wise man, who simultaneously protects the 'Castle of the Clouds' and carries gifts metamorphosis energy as *spirit mercurius* (*Anima Mercurii*).

The appearance of this image marked a point of change in the analytical process, which only became noticeable after some time, when Mr. M. began to assume his role as father. The drawing accompanied us from then on, and I could repeatedly convey my initially silent amplification of Mercurius to apparently contradictory events in the mind and feelings of the analysand. His inherent ambivalences corresponded to the fact that Mercurius is actually a duality: he consists of all conceivable opposites. He is physical and spiritual at the same time. He describes the process of transformation of the lower, physical into the upper, spiritual and vice versa. He is both the devil and a pioneering saviour. He is a trickster and a deity of nature. Mercurius manifests the mystical experience of an inner change. On the one hand he portrays the Self, on the other hand the process of individuation and furthermore in his boundlessness also the collective unconscious (Jung, 1942, GW 13; Dieckmann, 1984, p. 55ff.).

Before hearing about his girlfriend's second pregnancy, the following change began: Mr. M. became happier, reported progress in his drawing and sketched a children's story in ten pictures about a maggot living alone in a positive motherly pear on a tree, isolated and threatened by all things surrounding it.

In the picture story, the defenceless maggot is threatened by mortal danger. Being eaten by either the dangerous mythological *Father* animal 'Black Bird' or the negative-holding mythological *Mother* animal 'Spider' seems certain and inevitable (see Figure 10.5).

Rescue comes from a force of nature: a thunderstorm throws the pear off the tree. The maggot, now 'grounded', very happily encounters an earthworm on the ground in the grass between flowers – it is one of its kind, alter ego, with which it can move into the distance (see Figure 10.6).

With the image of the worm, the 'snake' has been revived, symbolizing a mythological animal in the Mother archetype. Like the Paradise Snake it calls for exploration, knowledge and progression. The appearance of the friendly 'worm/snake/dragon' is significant here in that it symbolically occurs at a turning point in this therapy, from which the analysand embarks on a forward moving journey.

Nine months later he draws an important dream picture (Figure 10.7).

A golden glow is seen from a gate and four inviting female figures without faces stand right and left; he had the feeling that if he were to walk through the door, he would fall abysmally. His pursing thought made him determined not to walk through the gate: '*I don't want to die yet.*' The quaternal

Figure 10.5 Threatening animal symbols in the parent complexes.

Figure 10.6 Helpful animal symbols.

Figure 10.7 The women's gate.

is a significant symbol of wholeness. On the one hand, he had the tendency to walk through the gate, to let himself fall into abyss, to experience the unconscious and no longer to meet it rationally – on the other hand: he was still governed by too much fear. An eerie stimulation of the psychological atmosphere through 'succubi', female demons, has taken place (Jung, 1943, GW 12, § 59): The enticement of women can also be sirens or *lamiens* that bewitch and mislead. In this sense, the gate could also be the stage for a Jimi Hendrix like performance, posing as symbol of the temptation of a fantasy of grandiosity that leads to fall and death.

I further associated a birth canal with this picture. I interpreted accordingly, whereupon he for the first time expressed clearly that he had been an unwanted child. His birth was a torture for mother, leaving him with deformed head. Therefore, the emergence of these archetypal motives initiated the removal of the inhibitory stagnation in his individuation process. This archetypal dream symbolism initiated a development.

He did not appear to our next session: his son had been born healthy and wonderful, and he had forgotten the session in his excitement. In his birth announcement, mother and child appeared friendly and giant facing a small, curious little mouse. The wider environment is marked by demonic, noisy and dirty chaos.

Chapter 11

Farewell to life

Mr. M became calmer and increasingly balanced. He began to make professional arrangements that were grounded and realistic. Repeatedly, we have been working on issues of dependence on women and his mother and the outstanding wishes to the father.

He came across a *'representative image of my former situation'* in a magazine: a young boy is threatened to be pulled under water by a climbing plant and a mermaid. He screams for help in utmost distress, but in vain, since there are but other weak children. His comment: *'only if the boy remains calm he can safe himself.'* His own inclination for panic could be talked about and understood as part of his 'bodily wisdom' – his fear deemed as a meaningful and useful reaction.

Together with his girlfriend, he has been visiting his mother. He considered her formal and inept in handling the child and was astounded by the observation that she seemed uninterested in the child beyond the formal grandmotherly demeanour: *'apparently she wanted to be left in peace'*. Following this insight, he could become more accepting of his mother and his inner image of his parents' relationship had changed: *'mother and father have probably given their best.'* This development can be understood as *humanization* of the imagines of parents and of a family myth (Kradin, 2009, p. 224ff.) that was driven by overly powerful parents' severity and expectations to excel.

The dream image of the four female figures became of renewed interest: he believed he could now cross through the gate. So far, this imagination seems always to have dispersed his *structures of thinking*. He realized that he had accounted all former and current female friends as well as his mother belonging to an *inner haram*.

Finding a company looking for an art director in a newspaper ad, he decided to apply *'without regard to any formalities: I want to scream out who I am, show my true colours, take a stand'*. Even though he received a negative response and spent his birthday in a gloomy mood, he received a call two days later: there has been a mistake; he could have the job. Mr. M. has been working there successfully to date as a valued colleague.

After he was hired, he dreamt:

> I am in a room. Suddenly a plastic helicopter flies in through an open window. Grateful for this opportunity I walk up to the window to see where it came from. Suddenly two drunk teenagers coming running in through the terrace door to shortcut through my apartment. I overwhelm them and beat them and throw them out through my door. One screams: You're just jealous of our freedom. This affects me and leaves me behind feeling thoughtful.

Joining the 'wild young boys', entrapped in a *Puer*-attitude, has not been possible any longer, even if the 'freedom' was tempting. The dream ego of Mr. M. proved himself 'king of his castle', who would not be fooled by the drunken and destructive emotions – using force if needed to sort out.

In accordance, his clothing has begun to change around this time. He now wore a suit jacket and fancy shirts and trousers. His university notes were binned. While his girlfriend attended a week-long professional training course, he took care of their son alone, experiencing an intensive and mutual bond.

Before the final analytic session, he impulsively felt like *eel and a glass of champagne* and got himself both. He remembered: eel and champagne was served to him by his uncle when he first left home. His uncle had asked him at the time: *'Are you gonna make it?'* He showed up to our final session with tonsillitis but happy: *'I guess I have achieved something – I feel like a child crossing the street alone for the first time.'*

He brought this final departure dream:

> I banish you and this therapy to the edge of the world.

Summary – course of treatment

The analysand, 41 years old at the end of treatment, begun psychotherapy at a stage in life when his girlfriend became pregnant and fear crippled him at the thought of having to take responsibility for becoming a father, after years of having felt trapped in a sub-depressive and regressive in-between state. Consequently, severe depressive mood with acute suicidal ideation arose.

Unconsciously, he had been identified with a particularly weak mother imago, which nonetheless threatened to hinder any development by way of a powerful, retentive mother complex (paired with a negative *elementary character* according to Neumann, 1974, p. 147ff.).

His relating to emotions, his ability to love, his relationship to his own unconscious and therefrom his transformative potentials were regressively impressed, leading to perpetual internal devaluation culminating in his death wish.

His father imago was coloured by heroic as well as extremely fearsome elements. In his father complex collective components weighed heavy: an upbringing to autonomy and the allowance to take responsibility never arose.

The compulsive neurotic side of the analysand worked ambivalently in the process: while it helped him to maintain the analysis and at his position at work, it also became noticeable as a tenacious resistance to change during a prolonged treatment phase. As Mr. M began to acknowledge the splitting off his feelings from his experiences, compulsive neurotic attitudes weakened. Primarily, he perceived this dynamic as to his benevolent feelings. Only towards the end of the treatment, aggressive elements concerning our relation surfaced, which have been spared from realization in order not to harm our relationship.

Symptomatically, he improved significantly with regard to his disorientation, naughtiness and depressiveness.

Typologically, the feeling function remained the 'slow function', even if attitudinal changes occurred in improved perception of *feeling*, and *thinking* had also become much more related and structured. His *intuition* underscored development: both in his professional development as well as in his choice of partner, there was a clear, unimagined *leitmotif*. The possibility of introversion had also developed in the perception and understanding of dreams and inner images.

A major part of the investigation of the unconscious via inner and outer symbolic images occurred through the predominantly visually structured main function, *sensation*. Symbolic images were beset with emotions and acted compensatory. His *transcendent function* grew stronger on the basis of these images by means of constructive and consciously perceivable meaning connected to relevant unconscious content.

I understand uncoupling and individuation processes in terms of an energetic weakening of effective pathogenic complexes vis-à-vis the ego complex. The ego complex has initially been identified with the image of the 'eternal youngling' (*Puer aeternus*). This pubertal, immature demeanour could be increasingly lifted by his relationship to his girlfriend and his son. Correspondingly, his narcissistic demeanour weakened significantly.

The symbolic unfolding and activation of the ego complex ensued step by step through the being-inside-the-egg towards the maggot in the apple, from there to the young boy, who remained threatened by the undine-like mermaid, but who could keep himself if he were to remain calm – which he succeeded by adopting a father's point of view.

I experienced the encounter with the analysand as an intersubjective and dialogical process of understanding and relationship building. Through my containing and interpreting functions, the aforementioned dreams became acceptable. Apart from the emotional and cognitive process of understanding, the creation of a *third space* intended for the search for meaning aided crucial steps in the analysand's individuation towards an extension of his abilities to love, to work and *doing* relationship. The painfulness of saying good-bye and the impossibility of realizing his desire to stay had to culminate in the need to banish the analyst and psychotherapy *to the end of the world.*

Health insurance benefits and therapy termination: as much as possible or as much as necessary?

The analysand ended the analytical psychotherapy with 20 insurance-covered sessions left. He argued that he needed to fully concentrate on his new job and live his life now. Thus, he was able to end his analytical treatment with an autonomous decision, even if some issues remained unresolved in his view.

Our case example ended in a shared process and signified a changing point in life. Apart from his responsibility for his wife and child, his new job became so important that he could tell himself and me that he needs no further sessions, and he is not in sufficient need to justify any more meetings any longer.

We often forget the worldwide uniqueness of the generous health insurance coverage for psychotherapy in Germany. Review procedure grants a substantial amount of treatment sessions with regard to the indicated and chosen treatment type. Therefore, to keep in mind the necessity of ongoing session as well as the appropriate moment of ending treatment is crucial in terms of both content and ethics.

It is important and useful for *both* the analysand and the analyst to take joint responsibility explicitly and right from the beginning for the timely framework and the end point of the treatment process. Accordingly, it is the analysand who applies for cost coverage with his health insurance, while the analyst drafts the report to be reviewed by an expert justifying both indication and the expected scope of sessions. Patient's rights include the right of access to the notes and diagnostic findings of the analyst as well as right of access to the review reports to the health insurance's experts. The report ought to be drafted in a way that can be understood not only by an expert colleague but also by the analysand. The analysand should be able to understand the report as an expression of therapeutic relationship, in which his most intimate and important issues are preserved, but into a psychodynamic context, safekept and guarded.

Healing dimensions in individuation

Since ancient times a philosophical path, alongside a religious one, promised inner peace and liberation, namely *self-knowledge – gnothi seauton, nosce te ipsum, tat tvam asi.*

Our case example highlights the goal of self-acceptance imminent in the analytical process and has been exemplified particularly in relation to the shadow, psychological complexes and goals of individuation. The latter is closely linked to our narcissistic needs for recognition, acceptance and admiration by others. A demeanour that allows for open and honest critique by others to gain self-insight is difficult to venture and develop. A chiefly relaxed, even jovial, modest self-determination as an expression of one's own identity, in which pain and suffering can be experienced and accepted, appears even more precarious. Encountering a wisely individuated person seems rare. Jungian psychotherapy and psychoanalysis inherently seeks to facilitate an analysand's access onto this path.

An essential premise of Jungian psychotherapy and psychoanalysis is its predominantly optimistic future outlook, manifest in a lifelong individuation path. This path is fuelled by the inner necessity of effective symbols of the unconscious that become implemented in interaction with other persons.

Sigmund Freud sounded considerably more pessimistic when he speaks of the lack of creation's intention for humans to be happy in his late work *Civilization and Its Discontents* (1929, GW 14, S. 434f.). He elaborates:

> Happiness strictly speaking arises more so of a sudden satisfaction of heavily accumulated needs and is naturally only possible as an episodic phenomenon. Any enduring of a longed-for situation underlain by the pleasure principle only results in a feeling of lukewarm contentment. We are equipped to enjoy intensively only what is contrast and much less so what is an outlasting state. Therefore, our possibility for happiness is restricted by our constitution. Unhappiness on the other hand is experienced with much more ease. Three ways offer passageway for suffering. Our own body, which, doomed for forfeiture and disintegration, cannot dispense with warning signals of pain and fear. The outside world with its overpowering, relentless and destructive forces that can ravage against us. And finally our relationships with other humans. Suffering protruding from the latter source might be what appears to us most agonizing. We seem to interpret it as an unnecessary ingredient, albeit it appears not less fateful that any of the other sources of suffering.

In contrast, the intersubjective perspective of all Freudian, Kleinian or Jungian psychoanalysis informed by early infant research advances a view of happiness surmising the idea that a sufficiently satisfactory relationship to others and to one's own creativity exists.

Many analysands hope to find a way back into *wanting to connect* despite the current critical introversion and fear of contact and notwithstanding the inevitable conflicts brought about by human contact. To accept the strangeness and difference of the other, our willingness to empathize nonetheless and the ability to at least temporarily view *us* as a 'couple' all make us more able to *do* relationship, care, love and work.

In the theory of Jungian psychotherapy and psychoanalysis, the death drive is not hidden 'beyond the pleasure principle' but rather the idea of the fundamental connection of life with nature as a whole (*unus mundus*). Certain qualities of human encounter, earlier described as 'now moments' – enamouring, enchanting spiritual touches or synchronistic phenomena – apparently arise out of a fundamental or *psychoid* connection of all being. Therein, the psychic, the archetypical of the shared unconscious and the possibility of symbolic connection, is set out to be.

On an important final note, we – both as analysands and analysts – unescapably confront all possibilities of daily experience of human perfidy, collective aggression and interpersonal iciness and cruelty ongoingly or, as Freud said, *fatefully*. A psychoanalytic experience may spark a venture of an upright personal stride as much as an active ethnical relationship to the hybridity of our baffling, brittle and media-billowed world.

Bibliography

Altmeyer, M. (2000). *Narzissmus und Objekt. Ein intersubjektives Verständnis der Selbstbezogenheit.* Göttingen: Vandenhoeck & Ruprecht.

Arbeitskreis OPD (Hrsg.) (2009). *Operationalisierte Psychodynamische Diagnostik OPD-2. Das Manual für Diagnostik und Therapieplanung.* Bern: Hans Huber.

Asper, K. (1989). *Verlassenheit und Selbstentfremdung.* Olten: Walter.

Baars, B. J. (2005). Global workspace theory of consciousness: Towards a cognitive neuroscience of human experience? *Progress in Brain Research, 150,* 45–54.

Bahner, E. (2002). Moderne Mythen – Autopoiese und Intersubjektivität. *Analytische Psychologie, 33,* 206–220.

Beebe, J. (2010). The recognition of psychological type. In M. Stein (Ed.), *Jungian Psychoanalysis. Working in the Spirit of C. G. Jung* (pp. 71–80). Chicago, La Salle: Open Court.

Benjamin, J. (1996). *Phantasie und Geschlecht. Psychoanalytische Studien über Idealisierung, Anerkennung und Differenz.* Frankfurt/M.: S. Fischer.

Bettighofer, S. (2016). *Übertragung und Gegenübertragung im therapeutischen Prozess.* Stuttgart: Kohlhammer.

Bion, W. R. (1997). *Lernen durch Erfahrung.* Frankfurt/M.: Suhrkamp.

Bion, W. R. (2007). *Die Tavistock-Seminare.* Tübingen: edition discord.

Blomeyer, R. (1979). Zur Theorie und Praxis typologischer Zuordnungen. Wertungen unter Analytikern. *Analytische Psychologie, 10,* 103–127.

Blomeyer, R. (1988). Anmerkungen zur Typologie. *Analytische Psychologie, 19,* 98–127.

Blomeyer, R. (1995). Über Interpretationen. *Analytische Psychologie, 26,* 75–78.

Bollas, C. (1997). *Der Schatten des Objekts. Das ungedachte Bekannte: Zur Psychoanalyse der frühen Entwicklung.* Stuttgart: Klett-Cotta.

Bollas, C. (2000). *Genese der Persönlichkeit.* Stuttgart: Klett-Cotta.

Botella, C. & Botella, S. (2005). *The Work of Psychic Figurability.* New York: Brunner-Routledge.

Bovensiepen, G. (2004). Bindung-Dissoziation-Netzwerk. Überlegungen zur Komplextheorie vor dem Hintergrund der Säuglingsforschung und der Neurowissenschaft. *Analytische Psychologie, 135,* 31–53.

Bovensiepen, G. (2009). „Leben in der Seifenblase". *Analytische Psychologie, 156,* 135–151.

Bovensiepen, G. (2011). C. G. Jung heute – der analytische Prozess. *Analytische Psychologie, 165,* 288–312.

Bovensiepen, G. (2019). *Die Komplextheorie. Ihre Weiterentwicklungen und Anwendungen in der Psychotherapie*. Stuttgart: Kohlhammer.

Braun, C. (2004). Der Mythos der introvertierten Individuation. *Analytische Psychologie, 135*, 423–447.

Braun, C. (2005). Individuation in Psychoanalysen – ein dialogischer Prozess? In L. Otscheret & C. Braun (Hrsg.), *Im Dialog mit dem Anderen. Intersubjektivität in Psychoanalyse und Psychotherapie* (S. 85–116). Frankfurt/M.: Brandes & Apsel.

Braun, C. (2010a). Editorial. *Analytische Psychologie, 162*, 393–397.

Braun, C. (2010b). Individuation und Träume. *Analytische Psychologie, 162*, 445–457.

Braun, C. & Otscheret, L. (2004). *Sexualitäten in der Psychoanalyse*. Frankfurt/M.: Brandes & Apsel.

Braun, C. & Otscheret, L. (2010). Dialogue. In M. Stein (Ed.), *Jungian Psychoanalysis. Working in the Spirit of C. G. Jung* (pp. 179–187). Chicago, La Salle: Open Court.

Britton, R. (2006). *Sexualität, Tod und Über-Ich*. Stuttgart: Klett-Cotta.

Buber, M. (1923). *Das dialogische Prinzip*. Heidelberg: Lambert Schneider.

Buchholz, M. B. (2003). *DGPT-Psycho-Newsletter 13*, Typoskript.

Buchholz, M. B. & Kleist, C. v. (1997). *Szenarien des Kontakts. Eine metaphernanalytische Untersuchung stationärer Psychotherapie*. Gießen: Psychosozial-Verlag.

Cambray, J. (2001). Enactments and amplification. *Journal of Analytical Psychology, 46*, 275–303.

Cambray, J. & Carter, L. (Eds.) (2004). *Analytical Psychology. Contemporary Perspectives in Jungian Analysis*. Hove, New York: Brunner-Routledge.

Cambray, J. & Carter, L. (2004a). Analytical methods revisited. In J. Cambray & L. Carter (Hrsg.), *Analytical Psychology. Contemporary Perspectives in Jungian Analysis* (pp. 116–148). Hove, New York: Brunner-Routledge.

Canestri, J. (2001). Feuerlärm: Überlegungen zur Übertragungsliebe. In E. S. Person, A. Hagelin & P. Fonagy (Hrsg.), *Über Freuds 'Bemerkungen über die Übertragungsliebe'* (S. 183–204). Stuttgart – Bad Cannstatt: Fromann-Holzbog.

Carter, L. (2010). The transcendent function, moments of meeting and dyadic consciousness: constructive and destructive co-creation in the analytic dyad. *Journal of Analytical Psychology, 55*, 217–227.

Chamisso, A. v. (1923). *Peter Schlehmils wundersame Geschichte*. München: Deutsche Meister-Verlag.

Crowther, C. & Schmidt, M. (2015). States of grace. *Journal of Analytical Psychology, 60*, 54–74.

Deserno, H. (1992). Traum und Übertragung. *Psyche, 10*, 959–978.

Deserno, H. (Hrsg.) (1999). *Das Jahrhundert der Traumdeutung*. Stuttgart: Klett-Cotta.

Deserno, H. (1999a). Der Traum im Verhältnis Übertragung und Erinnerung. In H. Deserno (Hrsg.), *Das Jahrhundert der Traumdeutung* (S. 397–431). Stuttgart: Klett-Cotta.

Dieckmann, H. (1979). *Methoden der analytischen Psychologie. Eine Einführung*. Olten: Walter.

Dieckmann, H. (1983). Individuelle und kollektive Wandlungschancen durch den Traum. *Analytische Psychologie, 14*, 50–64.

Dieckmann, H. (1984). *Träume als Sprache der Seele*. Fellbach: Adolf Bonz.

Dieckmann, H. (1991a). *Komplexe*. Berlin, Heidelberg, New York: Springer.

Dieckmann, H. (1991b). Dogma und freier Geist. Überlegungen zur Systematik der Analytischen Psychologie C. G. Jungs. *Analytische Psychologie, 22,* 157–173.

Dolto, F. (1996). *Über das Begehren.* Stuttgart: Klett-Cotta.

Döring, N. (2003). *Sozialpsychologie des Internet.* Göttingen, Bern, Toronto, Seattle: Hogrefe.

Dornes, M. (2000). *Die emotionale Welt des Kindes.* Frankfurt/M.: Fischer.

Dornes, M. (2010). Die Modernisierung der Seele. *Psyche, 11 (64),* 995–1033.

Dorst, B. (2015). *Therapeutisches Arbeiten mit Symbolen.* Stuttgart: W. Kohlhammer.

Ehrenberg, A. (2008). *Das erschöpfte Selbst. Depression und Gesellschaft in der Gegenwart.* Frankfurt/M.: Suhrkamp.

Erikson, E. H. (1998). *Der vollständige Lebenszyklus.* Frankfurt/M.: Suhrkamp.

Eschenbach, U. (Hrsg.) (1983). *Die Behandlung in der Analytischen Psychologie III.* Fellbach: Adolf Bonz.

Fairbairn, W. R. D. (1952/2000). *Das Selbst und die inneren Objektbeziehungen.* Gießen: Psychosozial-Verlag.

Ferro, A. (2003). *Das bipersonale Feld.* Gießen: Psychosozial-Verlag.

Fonagy, P. & Target, M. (2006). *Psychoanalyse und die Psychopathologie der Entwicklung.* Stuttgart: Klett-Cotta.Fordham, M. (1974). Notes on the transference. In M. Fordham, R. Gordon, J. Hubback & K. Lambert (Eds.), *Technique in Jungian Analysis* (pp. 147–160). London, New York: Karnac.

Fordham, M. (1985). *Explorations into the Self.* London, New York: Karnac.

Fordham, M., Gordon, R., Hubback, J. & Lambert, K. (Eds.) (1974). *Technique in Jungian Analysis.* London, New York: Karnac.

Franz, M.-L. v. (1987). *Der ewige Jüngling.* München: Pfeiffer.

Franz, M.-L. v. & Hillman, J. (1980). *Zur Typologie C. G. Jungs.* Fellbach-Öffingen: Adolf Bonz.

Freud, S. (1915). Bemerkungen über die Übertragungsliebe. *GW 10,* S. 306–321.

Freud, S. (1927). Der Humor. *GW 14,* S. 381–389.

Freud, S. (1929). Das Unbehagen in der Kultur. *GW 14,* S. 419–506.

Freud, S. (1981). Gesammelte Werke (GW) 1–18, 7. Aufl. London: Imago Publishing.

Fröbe-Kaptein, O. (Hrsg.). *Eranos-Jahrbuch 1957.* Zürich: Rhein-Verlag.

Fromm, E. (1936). *Sozialpsychologischer Teil der Studien über Autorität und Familie. Forschungsberichte aus dem Institut für Sozialforschung.* Paris: Librairie Félix Alcan.

Fuchs, T. (2011). Gehirnkrankheiten oder Beziehungsstörungen? Eine systemisch-ökologische Konzeption psychischer Krankheit. In G. Schiepek (Hrsg.), *Neurobiologie der Psychotherapie* (S. 375–383). Stuttgart: Schattauer.

Gallese, V. (2015). Welche Neurowissenschaften und welche Psychoanalyse? *Psyche, 69,* 97–114.

Giegerich, W. (2012). *What Is Soul?* New Orleans, Louisiana: Spring Journal Books.

Giegerich, W. (2014). Neurose: das Werk der kranken Seele. *Analytische Psychologie, 45, 177,* 288–300.

Goethe, J. W. v. (1998). *Werke. Hamburger Ausgabe in 14 Bänden. Band 1.* München, Deutscher Taschenbuch Verlag.

Grimm, J. & Grimm, W. (Hrsg.) (1985). *Die Kinder- und Hausmärchen der Brüder Grimm in ihrer Urgestalt.* Lindau: Antiqua-Verlag.

Guderian, C. (2007). *Die Couch in der Psychoanalyse. Geschichte und Gegenwart von Setting und Raum.* Stuttgart: W. Kohlhammer.

Guggenbühl-Craig, A. (1976). Psychopathia sexualis und Jungsche Psychologie. *Analytische Psychologie, 7,* 110–122.

Habermas, J. (2005). *Zwischen Naturalismus und Religion.* Frankfurt a.M.: Suhrkamp.

Heimann, P. (1950). On countertransference. *International Journal Psycho-Analysis, 31,* 81–84.

Hill, J. (2010). Amplification: Unveiling emergent patterns of meaning. In M. Stein (Ed.), *Jungian Psychoanalysis. Working in the Spirit of C.G. Jung* (pp. 109–117). Chicago, La Salle: Open Court.

Höhfeld, K. (1997). Individuation und Neurose. *Analytische Psychologie, 28,* 188–202.

Hogenson, G. B. (2004). What are symbols of? *Journal of Analytical Psychology, 49,* 67–81.

Hogenson, G. B. (2005). The self, the symbolic and synchronicity. *Journal of Analytical Psychology, 50,* 271–284.

Hohage, R. (2001). *Analytische und tiefenpsychologisch fundierte Psychotherapie: Unterschiede zwischen analytischer und therapeutischer Haltung.* DPV-Tagung Herbst 2001, Vortragsmanuskript.

Honneth, A. (2002). Objektbeziehungstheorie und postmoderne Identität. Über das vermeintliche Veralten der Psychoanalyse. *Psyche, 54,* 1087–1108.

Jacoby, M. (1975). Autorität und Revolte – der Mythos vom Vatermord. *Analytische Psychologie, 6,* 524–540.

Jacoby, M. (1985). *Individuation und Narzissmus.* München: Pfeiffer.

Jacoby, M. (1993). *Übertragung und Beziehung in der jungschen Praxis.* Düsseldorf: Walter.

Jacoby, M. (1998). *Grundformen seelischer Austauschprozesse.* Zürich, Düsseldorf: Walter.

Jacoby, M. (2005). Zu den Wurzeln intersubjektive Bedürfnisse. In L. Otscheret & C. Braun (Hrsg.), *Im Dialog mit dem Anderen. Intersubjektivität in Psychoanalyse und Psychotherapie* (S. 14–28). Frankfurt/M.: Brandes & Apsel.

Jaffé, A. (Hrsg.) (1984). *Erinnerungen, Träume, Gedanken von C. G. Jung.* Olten, Freiburg: Walter.

Johnson, M. (1987). *The Body in the Mind. The Bodily Basis of Meaning, Imagination and Reason.* Chicago, London: University Press.

Jung, E. (1971): Der Großinquisitor. *Analytische Psychologie, 2,* 79–104.

Jung, E. (1974). Zur gleichgeschlechtlichen Übertragungs-Gegenübertragungskonstellation unter Berücksichtigung der Arileus-Vision. *Analytische Psychologie, 5,* 204–224.

Jung, C. G. (1907). Über die Psychologie der Dementia Praecox: ein Versuch. *GW 3,* §§ 1–76.

Jung, C. G. (1911). Symbole der Wandlung. Analyse des Vorspiels zu einer Schizophrenie. *GW 5,* §§ 1–685.

Jung, C. G. (1911a). Das Typenproblem in der Dichtkunst. *GW 6,* §§ 261–526.

Jung, C. G. (1913). Versuch einer Darstellung der Psychoanalytischen Theorie. *GW 4,* §§ 203–522.

Jung, C. G. (1914). Nachtrag: Über das psychologische Verständnis pathologischer Vorgänge. *GW 3,* §§ 388–424.

Jung, C. G. (1916). Die transzendente Funktion. *GW 8,* § 131–193.

Jung, C. G. (1916a). Anpassung, Individuation und Kollektivität. *GW 18/II,* §§ 1084–1109.

Jung, C. G. (1920). Die Beziehungen zwischen dem Ich und dem Unbewussten. *GW 7*, §§ 202–406.

Jung, C. G. (1921). Psychologische Typen. XI Definitionen. *GW 6*, §§ 741–921.

Jung, C. G. (1928). Über die Energetik der Seele. *GW 8*, §§ 1–130.

Jung, C. G. (1929). Die Probleme der modernen Psychotherapie. *GW 16*, §§ 114–174.

Jung, C. G. (1930). Die Lebenswende. *GW 8*, §§ 749–795.

Jung, C. G. (1930a). Einführung zu W. M. Kranefeldt „Die Psychoanalyse". *GW 4*, §§ 745–767.

Jung, C. G. (1931). Die praktische Verwendbarkeit der Traumanalyse. *GW 16*, §§ 294–352.

Jung, C. G. (1931a). Einführung zu Frances G. Wickes „Analyse der Kinderseele". *GW 17*, §§ 80–97.

Jung, C. G. (1933). Über die Archetypen des kollektiven Unbewussten. *GW 9/I*, §§ 1–86.

Jung, C. G. (1934). Allgemeines zur Komplextheorie. *GW 8*, §§ 194–219.

Jung, C. G. (1934a). Zur gegenwärtigen Lage der Psychotherapie. *GW 10*, §§ 333–370.

Jung, C. G. (1940). Zur Psychologie des Kindarchetypus. *GW 9/I*, §§ 259–305.

Jung, C. G. (1942). Der Geist Mercurius. *GW 13*, §§ 239–303.

Jung, C. G. (1943). Traumsymbole des Individuationsprozesses. *GW 12*, §§ 43–331.

Jung, C. G. (1944). Einleitung in die religionspsychologische Problematik der Alchemie. *GW 12*, §§ 1–42.

Jung, C. G. (1945). Die Psychologie der Übertragung. *GW 16*, §§ 353–539.

Jung, C. G. (1946). Theoretische Überlegungen zum Wesen des Psychischen. *GW 8*, §§ 343–442.

Jung, C. G. (1954). Der Philosophische Baum. *GW 13*, §§ 304–482.

Jung, C. G. (1961). Symbole und Traumdeutung. *GW 18/I*, §§ 416–607.

Jung, C. G. (1976–1985). *Gesammelte Werke in 20 Bänden* (GW 1–20), Olten, Freiburg: Walter.

Jung, C. G. (1981). *Briefe I 1906–1945*. Olten: Walter.

Jung, C. G. (1989): *Briefe II 1946–1955*. Olten: Walter.

Jung, C. G. (2009). *Das Rote Buch. Liber novus*. Düsseldorf: Patmos.

Jung, C. (2005). „Der erste Gegenstand des Menschen ist der Mensch" – Ludwig Feuerbach entdeckte die Dialogik. In L. Otscheret & C. Braun (Hrsg.), *Im Dialog mit dem Anderen. Intersubjektivität in Psychoanalyse und Psychotherapie* (S. 216–235). Frankfurt/M.: Brandes & Apsel.

Kächele, H. (2015). *Das Wunder ist des Glaubens liebstes Kind*. Zugriff am 20.12.2015 unter: www.uniklinik-ulm.de/fileadmin/Kliniken/Psych_Medizin_Psychtherapie/Dokumente/Lehre/placebo.pdf.

Kaplan-Solms, K. & Solms, M. (2000). *Neuro-Psychoanalyse*. Stuttgart: Klett-Cotta.

Kast, V. (2006). *Träume*. Düsseldorf: Patmos, Walter.

Kast, V. (2010). Träume als Wegweiser. *Analytische Psychologie, 162*, 420–431.

Kerényi, K. (1997). *Die Mythologie der Griechen II. Die Heroen-Geschichten*. Stuttgart: Klett-Cotta.

Keupp, H. (2012). Identität und Individualisierung. In H. G. Petzold (Hrsg.), *Identität* (S. 77–105). Wiesbaden: VS-Verlag.

Keupp, H. & Höfer, R. (Hrsg.) (1997). *Identitätsarbeit heute*. Frankfurt/M.: Suhrkamp.

Kirsch, T. B. (2007). *C. G. Jung und seine Nachfolger*. Gießen: Psychosozial-Verlag.

Klein, M. (1972a). *Das Seelenleben des Kleinkindes und andere Beiträge zur Psychoanalyse*. Hamburg: Rowohlt.

Klein, M. (1972b). Über das Seelenleben des Kleinkindes. In M. Klein (Ed.), *Das Seelenleben des Kleinkindes und andere Beiträge zur Psychoanalyse* (S. 144–173). Hamburg: Rowohlt.

Klingenberg-Vogel, M. (2008). „Was ist Liebe?" – Wie kann ein Patient „liebesfähig" werden? In A. Springer, K. Münch & D. Munz (Hrsg.), *Sexualitäten* (S. 181–198). Gießen: Psychosozial-Verlag.

Knox, J. (2004). Developmental aspects of analytical psychology: New perspectives from cognitive neuroscience and attachment theory. In J. Cambray & L. Carter (Eds.), *Analytical Psychology. Contemporary Perspectives in Jungian Analysis* (pp. 56–82). Hove, New York: Brunner-Routledge.

Knox, J. (2011). Die analytische Beziehung: eine Zusammenführung jungianischer, bindungstheoretischer und entwicklungspsychologischer Perspektiven. *Analytische Psychologie, 166*, 403–426.

Knox, J. (2012). Selbstwirksamkeit in Beziehungen. *Analytische Psychologie, 170, 43*, 450–471.

Körner, J. (2003). Die argumentationszugängliche Kasuistik. *Forum der Psychoanalyse, 19*, 28–35.

Körner, J. (2014). Arbeit „in" der Übertragung. *Forum der Psychoanalyse, 30*, 341–356.

Körner, J. & Krutzenbichler, S. (Hrsg.) (2000). *Der Traum in der Psychoanalyse.* Göttingen: Vandenhoeck & Ruprecht.

Kradin, R. (2009). The family myth: ids deconstruction and replacement with a balanced humanized narrative. *Journal of Analytical Psychology, 54*, 217–232.

Krappmann, L. (1997). Die Identitätsproblematik nach Erikson aus interaktionistischer Sicht. In H. Keupp & R. Höfer (Hrsg.), *Identitätsarbeit heute* (S. 66–92). Frankfurt/M.: Suhrkamp.

Kreuzer-Haustein, U. (2000). Der Traum in der Psychoanalyse. In J. Körner & S. Krutzenbichler (Hrsg.), *Der Traum in der Psychoanalyse* (S. 89–101). Göttingen: Vandenhoeck & Ruprecht.

Küchenhoff, J. (1999). Verlorenes Objekt, Trennung und Anerkennung. *Forum der Psychoanalyse, 15*, 189–203.

Lambert, K. (2002). *Analysis, Repair and Individuation.* London, New York: Karnak.

Langendorf, U. (1994). „Selbst ist der Mann" – Individuation als Ideal – Muss die Jungsche Theorie der Individuation revidiert werden? *Analytische Psychologie, 25*, 262–277.

Lesmeister, R. (2005). Technik und Beziehung. Erkundung eines Widerstreits. In L. Otscheret & C. Braun (Hrsg.), *Im Dialog mit dem Anderen. Intersubjektivität in Psychoanalyse und Psychotherapie* (S. 29–56). Frankfurt/M.: Brandes & Apsel.

Lesmeister, R. (2009). *Selbst und Individuation. Facetten von Subjektivität und Intersubjektivität in der Psychoanalyse.* Frankfurt/M.: Brandes & Apsel.

Lesmeister, R. (2011). C. G. Jung im psychoanalytischen Diskurs der Gegenwart. *Analytische Psychologie, 165*, 270–287.

Lloyd Mayer, E. (2002). Freud and Jung. *Journal of Analytical Psychology, 47*, 91–99.

Loewe, A. (2014). „Auf Seiten der inneren Stimme ...". Freiburg, München: Verlag Karl Albert.

Maier, C. (2014). Über die intersubjektive Entwicklung von Bedeutung. *Analytische Psychologie, 178*, 365–377.

Marlan, S. (2010). Facing the shadow. In M. Stein (Ed.), *Jungian Psychoanalysis. Working in the Spirit of C. G. Jung* (pp. 5–13). Chicago, La Salle: Open Court.

McFarland Solomon, H. (1997). Das gar nicht so schweigsame Paar im Individuum. *Analytische Psychologie, 28,* 149–171.

McFarland Solomon, E. (2010). The ethical attitude in analytical practice. In M. Stein (Ed.), *Jungian Psychoanalysis. Working in the Spirit of C. G. Jung* (pp. 325–333). Chicago, La Salle: Open Court.

McFarland Solomon, H. (2013). Wandlungspotenziale. *Analytische Psychologie, 172,* 216–251.

McGuire, W. & Sauerländer, W. (Hrsg.) (1974). *Sigmund Freud – C. G. Jung Briefwechsel.* Frankfurt/M.: S. Fischer.

Meier, I. (2015). Der klassische, gebannte und negative Held. *Analytische Psychologie, 179,* 8–26.

Meltzer, D. (2007). *Sexualität und psychische Struktur.* Tübingen: edition discord.

Mertens, W. (2002). Was ist von Freuds Traumpsychologie geblieben? In R. Zwiebel, M. Leuzinger-Bohleber (Hrsg.), *Träume, Spielräume I.* (S. 8–40). Göttingen: Vandenhoek & Ruprecht.

Mertens, W. (2003). *Traum und Traumdeutung.* München: Beck.

Mertens, W. (2015). *Psychoanalytische Behandlungstechnik.* Stuttgart: W. Kohlhammer.

Mizen, R. (2009). The embodied mind. *Journal of Analytical Psychology, 54,* 253–272.

Molino, J. (2000). Towards an evolutionary theory of music and language. In N. L. Wallin, B. Merker & S. Brown (Eds.), *The Origins of Music* (pp. 165–176). Cambridge, London: The MIT Press.

Morgenthaler, F. (1986). *Der Traum. Fragmente zur Theorie und Technik der Traumdeutung.* Frankfurt/M.: Campus.

Moser, U. & Zeppelin, I. v. (1996). *Der geträumte Traum. Wie Träume entstehen und sich ändern.* Stuttgart: W. Kohlhammer.

Moser, U. (2003). Traumtheorien und Traumkultur in der psychoanalytischen Praxis (Teil II). *Psyche, 57,* 729–750.

Neumann, E. (1957). Die Sinnfrage und das Individuum. In O. Fröbe-Kaptein (Hrsg.), *Eranos-Jahrbuch 1957* (S. 11–56). Zürich: Rhein-Verlag.

Neumann, E. (1974). *Die Große Mutter.* Olten: Walter.

Neumann, E. (1978). *Kulturentwicklung und Religion.* Frankfurt/M.: Fischer.

Neumann, E. (1999). *Ursprungsgeschichte des Bewusstseins.* Frankfurt/M.: Fischer.

Neumann, E. (2004). *Amor und Psyche. Eine tiefenpsychologische Deutung.* München: Patmos.

Odermatt, M. (1998). Individuation und gesellschaftliche Realität. *Analytische Psychologie, 29,* 256–271.

Ogden, T. H. (1994). *Subjects of Analysis.* Northvale, New York, London: Jason Aronson, Karnac.

Ogden, T. H. (2001). *Analytische Träumerei und Deutung. Zur Kunst der Psychoanalyse.* Wien, New York: Springer.

Orange, D. M. (2004). *Emotionales Verständnis und Intersubjektivität.* Frankfurt/Main: Brandes & Apsel.

Otscheret, L. (2005). Dialektik ohne Dialog. Intersubjektivität bei C. G. Jung. In L. Otscheret & C. Braun (Hrsg.), *Im Dialog mit dem Anderen. Intersubjektivität in Psychoanalyse und Psychotherapie* (S. 57–83). Frankfurt/M.: Brandes & Apsel.

Otscheret, L. (2010). Individuation und Veränderung in Traumserien. *Analytische Psychologie, 162,* 432–443.

Otscheret, L. & Braun, C. (Hrsg.) (2005). *Im Dialog mit dem Anderen. Intersubjektivität in Psychoanalyse und Psychotherapie.* Frankfurt/M.: Brandes & Apsel.

Panksepp, J. (1998). *Affective Neuroscience: The Foundations of Human and Animal Emotions.* New York: Oxford University Press.

Person, E. S. (1990). *Lust auf Liebe. Die Wiederentdeckung eines romantischen Gefühls.* Reinbek: Rowohlt.

Person, E. S. (1994). Die erotische Übertragung bei Frauen und Männern: Unterschiede und Folgen. *Psyche, 48,* 783–807.

Person, E. S., Hagelin, A. & Fonagy, P. (Hrsg.) (2001). *Über Freuds ,Bemerkungen über die Übertragungsliebe'.* Stuttgart – Bad Cannstatt: Fromann-Holzbog.

Petzold, H. G. (Hrsg.) (2012). *Identität.* Wiesbaden: VS-Verlag.

Plaut, F. (2004). *Between Losing and Finding. The Life of an Analyst.* London: Free Association Books.

Prinz, W. (2013). *Selbst im Spiegel. Die soziale Konstruktion von Subjektivität.* Berlin: Suhrkamp.

Quindeau, I. (2014). *Sexualität.* Gießen: Psychosozial-Verlag.

Racker, H. (1959/1988). *Übertragung und Gegenübertragung. Studien zur psychoanalytischen Technik.* München, Basel: Ernst Reinhardt Verlag.

Rafalski, M. (2011). Das individuelle Zusammenspiel der vier Orientierungsfunktionen. *Analytische Psychologie, 164,* 170–194.

Rass, E. (2011). *Bindung und Sicherheit im Lebenslauf.* Stuttgart: Klett-Cotta.

Ribi, A. (1983). Der Archetyp des Anthropos. In U. Eschenbach (Hrsg.), *Die Behandlung in der Analytischen Psychologie III* (S. 128–151). Fellbach: Adolf Bonz.

Ringelnatz, J. (1912). *Die Schnupftabaksdose.* E-Book Sammlung Zeno.org. Kindle Edition.

Rizzolatti, G. & Craighero, L. (2004). The 'Mirror-Neuron-System'. *Annual Review of Neuroscience, 27,* 169–192.

Rizzuto, A.-M. (2002). Response to questionnaire to practising psychoanalysts. *Journal of Analytical Psychology, 47 (1),* 7–15.

Roth, G. & Strüber, N. (2014). *Wie das Gehirn die Seele macht.* Stuttgart: Klett-Cotta.

Rudolph, G. (2005). *Strukturbezogene Psychotherapie.* Stuttgart, New York: Schattauer.

Samuels, A. (1989). *Jung und seine Nachfolger.* Stuttgart: Klett-Cotta.

Samuels, A. (2014). How to choose a therapist. zuletzt aufgerufen: 10.5.2016 http://welldoing.org/article/gender-race-age-what-matters-when-you-choose-a-therapist.

Samuels, A., Shorter, B. & Plaut, F. (1991). *Wörterbuch Jungscher Psychologie.* München: DTV.

Saunders, P. & Skar, P. (2001). Archetypes, complexes and self-organisation. *Journal of Analytical Psychology, 46,* 305–323.

Schiepek, G. (2011). *Neurobiologie der Psychotherapie.* Stuttgart: Schattauer.

Solms, M. & Turnbull, O. (2004). *Das Gehirn und die innere Welt.* Düsseldorf, Zürich: Walter.

Springer, A. (2000). Postjungianisches zum Traum in der Analysestunde. In J. Körner & S. Krutzenbichler (Hrsg.), *Der Traum in der Psychoanalyse* (S. 114–125). Göttingen: Vandenhoeck & Ruprecht.

Springer, A., Münch, K. & Munz, D. (Hrsg.) (2008). *Sexualitäten.* Gießen: Psychosozial-Verlag.

Staemmler, F.-M. (2015). *Das dialogische Selbst*. Stuttgart: Schattauer.

Stein, M. (Hrsg.) (2010). *Jungian Psychoanalysis. Working in the Spirit of C. G. Jung*. Chicago, La Salle: Open Court.

Stern, D. N. (1992). *Die Lebenserfahrung des Säuglings*. Stuttgart: Klett-Cotta.

Stern, D. N., Sander, L. W., Nahum, J. P., Harrison, A. M., Lyons-Ruth, K., Morgan, A. C., Bruschweiler-Stern, N. & Tronick, E. Z. (2002). „Nicht-deutende Mechanismen in der psychoanalytischen Therapie. Das ‚Etwas-Mehr' als Deutung". *Psyche, 56,* 974–1006.

Stern, D. N. (2005). *Der Gegenwartsmoment. Veränderungsprozesse in Psychoanalyse, Psychotherapie und Alltag*. Frankfurt/M.: Brandes & Apsel.

Treurniet, N. (1995). Was ist Psychoanalyse heute? *Psyche, 49,* 112–140.

Treurniet, N. (1996). Über eine Ethik der psychoanalytischen Technik. *Psyche, 50,* 1–31.

Trevarthen, C. (2010). What is it like to be a person who knows nothing? Defining the active intersubjective mind of a newborn human being. *The Intersubjective Newborn. Infant and Child Development, Special Issue, Edited by Emese Nagy*.

Vogel, R. (2008). *C. G. Jung für die Praxis*. Stuttgart: W. Kohlhammer.

Wallin, N. L., Merker, B. & Brown, S. (Eds.) (2000). *The Origins of Music*. Cambridge, London: The MIT Press.

Wehr, G. (2009). *Carl Gustav Jung. Leben – Werk – Wirkung*. Treuenbrietzen: Telesma.

Weil, G. (o. J.). *Tausend und eine Nacht. Arabische Erzählungen*. E-Book, Kindle-Edition.

Welsch, W. (2002). *Unsere postmoderne Moderne*. Berlin: Akademie.

Wiener, J. (2017, 2. Ed.). *The Therapeutic Relationship*. Texas: A&M University Press.

Wilkinson, M. (2004). The mind-brain relationship: The emergent self. *Journal of Analytical Psychology, 49,* 83–101.

Wilkinson, M. (2006). Die träumende Psyche – das träumende Gehirn. *Analytische Psychologie, 145,* 294–313.

Wilkinson, M. (2010). Psyche and brain. In M. Stein (Ed.), *Jungian Psychoanalysis. Working in the Spirit of C. G. Jung* (pp. 307–315). Chicago, La Salle: Open Court.

Winnicott, D. W. (1992). *Vom Spiel zur Kreativität*. Stuttgart: Klett-Cotta.

Winnicott, D. W. (2002). *Reifungsprozesse und fördernde Umwelt*. Gießen: Psychosozial-Verlag.

Zwiebel, R., Leuzinger-Bohleber, M. (Hrsg.) (2002). *Träume, Spielräume I*. Göttingen: Vandenhoek & Ruprecht.

Index

Note: *Italic* page numbers refer to figures.

For Product Safety Concerns and Information please contact our EU
representative GPSR@taylorandfrancis.com
Taylor & Francis Verlag GmbH, Kaufingerstraße 24, 80331 München, Germany

www.ingramcontent.com/pod-product-compliance
Lightning Source LLC
Chambersburg PA
CBHW050531270326
41926CB00015B/3179